# Application of Titanium Dioxide Photocatalysis to Construction Materials

# RILEM STATE-OF-THE-ART REPORTS

Volume 5

RILEM, The International Union of Laboratories and Experts in Construction Materials, Systems and Structures, founded in 1947, is a non-governmental scientific association whose goal is to contribute to progress in the construction sciences, techniques and industries, essentially by means of the communication it fosters between research and practice. RILEM's focus is on construction materials and their use in building and civil engineering structures, covering all phases of the building process from manufacture to use and recycling of materials. More information on RILEM and its previous publications can be found on www.RILEM.net.

The RILEM State-of-the-Art Reports (STAR) are produced by the Technical Committees. They represent one of the most important outputs that RILEM generates – high level scientific and engineering reports that provide cutting edge knowledge in a given field. The work of the TCs is one of RILEM's key functions.

Members of a TC are experts in their field and give their time freely to share their expertise. As a result, the broader scientific community benefits greatly from RILEM's activities.

RILEM's stated objective is to disseminate this information as widely as possible to the scientific community. RILEM therefore considers the STAR reports of its TCs as of highest importance, and encourages their publication whenever possible.

The information in this and similar reports is mostly pre-normative in the sense that it provides the underlying scientific fundamentals on which standards and codes of practice are based. Without such a solid scientific basis, construction practice will be less than efficient or economical.

It is RILEM's hope that this information will be of wide use to the scientific community.

For other titles published in this series, go to
www.springer.com/series/8780

Yoshihiko Ohama • Dionys Van Gemert
Editors

# Application of Titanium Dioxide Photocatalysis to Construction Materials

## State-of-the-Art Report of the RILEM Technical Committee 194-TDP

Springer

*Editors*
Yoshihiko Ohama
Nihon University
Hiyoshi 2-chome 14-10-402
223-0061 Kohoku-ku
Yokohama
Japan
ohama@bh.wakwak.com

Dionys Van Gemert
Civil Engineering Department
Katholieke Universiteit Leuven
Kasteelpark Arenberg 40
2001 Heverlee
Belgium
dionys.vangemert@bwk.kuleuven.be

ISBN 978-94-007-3609-2 ISBN 978-94-007-1297-3 (eBook)
DOI 10.1007/978-94-007-1297-3
Springer Dordrecht Heidelberg London New York



Printed on acid-free paper

Springer is part of Springer Science+Business Media (www.springer.com)

# Preface

This state-of-the-art report on "Application of Titanium Dioxide Photocatalysis to Construction Materials" was prepared by RILEM TC 194-TDP "Titanium Dioxide Photocatalysis". RILEM TC 194 was established in 2001, on proposition by Dr. Hiroyuki Yamanouchi, National RILEM Delegate for Japan. The aim of the TC 194 was to collect the theoretical data and practical achievements of titanium dioxide photocatalysis in construction. The beginning of practically applicable photocatalysis of $TiO_2$ originates with a discovery of the photocatalytic splitting of water on $TiO_2$ electrodes by two Japanese scientists, Fujishima and Honda in 1972 [15]. In the years after, the technology of $TiO_2$ photocatalysis has progressed rapidly, and became very attractive in the applications of self-cleaning, air-cleaning and antibacterial effects in the construction industry. The benefits of $TiO_2$ photocatalysis are further being exploited for cooling effects of heat island phenomena, and for water and soil treatments.

The committee members of RILEM TC 194-TDP have worked together for arranging this state-of-the-art report. Equilibrium was pursued between theoretical background information, practical application aspects in different construction areas, and standardization.

The committee held eight meetings, in which the different aspects of the state-of-the-art report were discussed, and the International Symposium on Photocatalysis, Environment and Construction Materials, held in Florence, 8–9 October 2007, was prepared [3]. The meetings took place in Tokyo (April 2002), Rome (September 2002), Lisbon (September 2003), Leipzig (April 2004), Koriyama (September 2004), Sterrebeek (October 2005), Madrid (May 2006) and London (May 2008).

The state-of-the-art report contains main chapters on principles of $TiO_2$ photocatalysis, on application of $TiO_2$ photocatalysis on cementitious materials for self-cleaning purposes, on application of antibacterial and self-cleaning effects to noncementitious construction materials, on applications of $TiO_2$ photocatalysis for air purification, and on standardization of testing methods for construction materials with $TiO_2$ photocatalysis. A list of selected references is presented as a guidance to the readers.

Prof. Dr. Y. Ohama
Chairman

Prof. Dr. D. Van Gemert
Secretary

# Authors and Acknowledgements

## Contributors – RILEM Technical Committee 194-TDP

RILEM gratefully acknowledges the dedication and contribution of time and effort by both committee members and external experts during the development of the work programme and the preparation of this state-of-the-art report on "Application of Titanium Dioxide Photocatalysis to Construction Materials". Sincere thanks are also expressed to the organisations, institutes and companies who provided the technical expertise through the generous support of their people who contributed to this report, and for their logistic support to the activities and work of the committee.

### TC Membership

**Chairman:**

**Yoshihiko Ohama**
Professor Emeritus
Nihon University, Tokyo, Japan

**Secretary:**

**Dionys Van Gemert**
Professor Emeritus
Katholieke Universiteit Leuven, Heverlee, Belgium

**Members:**

**Anne Beeldens**
Senior Researcher, Professor
Belgian Road Research Center, Belgium
and
Katholieke Universiteit Leuven, Heverlee, Belgium

**Detlef Bahnemann**
Professor
University of Hannover, Germany

**Luigi Cassar**
R&D Central Manager
CTG Italcementi Group, Italy

**Marta Castellote**
Senior Researcher
IETcc-CSIC, Spain

**Nele De Belie**
Professor
Ghent University, Belgium

**Frank Dehn**
Assistant Professor
University of Leipzig, Germany

**Keichi Imamoto**
Associate Professor
Tokyo University of Science, Japan

**Norimoto Kashino**
Professor
Maebashi Institute of Technology, Japan

**Makoto Kawakami**
Professor
Akita University, Japan

**July Maltby**
Researcher
Millennium Chemicals Europe, United Kingdom

**Anibal Maury Ramirez**
Researcher
Ghent University, Belgium

**Robert McIntyre**
Senior Researcher
Millennium Chemicals Europe, United Kingdom

**Kenji Motohashi**
Professor
Shibaura Institute of Technology, Japan

**Yoshihiko Murata**
Chief Research Engineer
Ube-Mitsubishi Cement Research Institute, Japan

**Yoshimitsu Saeki**
General Manager
Toto Ltd., Japan

**Tinne Vangheel**
Researcher
Belgian Building Research Centre, Belgium

**Andreas Wolf**
Senior Industry Scientist
Dow Corning, Germany

# Contents

**Preface** ... v

**Authors and Acknowledgements** ... vii

**1 Introduction** ... 1
Yoshihiko Ohama and Dionys Van Gemert

**2 Principles of $TiO_2$ Photocatalysis** ... 5
Marta Castellote and Nicklas Bengtsson
2.1 Definition of Photocatalysis ... 5
2.2 Oxidation of Organic Molecules by Photocatalysis ... 6
2.3 Other Chemical Concepts of Interest ... 7
2.4 Principles of $TiO_2$ Photocatalysis ... 7
2.4.1 Oxidation-Reduction Reactions ... 8
2.4.2 Development of Superhydrophilicity ... 9

**3 Application of $TiO_2$ Photocatalysis to Cementitious Materials for Self-Cleaning Purposes** ... 11
Anibal Maury Ramirez and Nele de Belie
3.1 Introduction ... 11
3.2 Self-Cleaning on Cementitious Materials ... 13

**4 Application of Antibacterial and Self-Cleaning Effects to Noncementitious Construction Materials** ... 17
Yoshimitsu Saeki
4.1 Introduction ... 17
4.2 How to Form Photocatalyst Films on Surfaces of Construction Materials ... 18
4.3 Functional Design and Supplement of Photocatalytic Effect in the Dark ... 19

4.3.1 Addition of Silica- or Silicon-Based Water-Retaining Substances .......... 19
4.3.2 Addition of Antibacterial Metals Such as Cu and Ag .......... 19
4.4 Photocatalysts Applied to Noncementitious Construction Materials .......... 20
4.4.1 Exterior Construction Materials .......... 20
4.4.2 Interior Construction Materials .......... 21

**5 Applications of $TiO_2$ Photocatalysis for Air Purification** .......... 23
Anne Beeldens, Luigi Cassar, and Yoshihiko Murata
5.1 The $TiO_2$-and-Cement System .......... 23
5.2 Air Purification Systems by Removing $NO_x$ .......... 24
5.3 Photocatalytic Blocks .......... 26
5.3.1 Feature of Photocatalytic Blocks .......... 26
5.3.2 $NO_x$ Removal Performance of Photocatalytic Blocks .......... 27
5.3.3 Examples of Applications .......... 30
5.4 Photocatalytic Pavements by Coating of Photocatalytic Polymer-Modified Pastes .......... 32

**6 Standardization of Testing Methods for Construction Materials with $TiO_2$ Photocatalyst** .......... 37
Kenji Motohashi, Frank Dehn, and Yoshihiko Ohama
6.1 Introduction .......... 37
6.2 The Existing Standards .......... 38

**7 Conclusions** .......... 43
Yoshihiko Ohama and Dionys Van Gemert

**References** .......... 45

# Chapter 1
# Introduction

**Yoshihiko Ohama and Dionys Van Gemert**

A photocatalyst is a semiconducting substance which can be chemically activated by light radiation that results in oxidation-reduction (redox) reaction, i.e., it has a photocatalysis action or photoactivity. As illustrated in Fig. 1.1 [16, p. 147; 24, p. 10], photocatalysis is similar to the photosynthesis in plants in the mechanism, in which chlorophyl acts as a catalyst to produce oxygen from carbon dioxide and water. Chlorophyl is a very powerful photocatalyst. The photocatalyst in the photocatalysis process corresponds to the chlorophyl in the photosynthesis process.

Many investigations to discover effective photocatalysts have been conducted up to the present time. However, titanium dioxide ($TiO_2$) with a crystal structure of anatase type has so far been found to be the most effective photocatalyst enabling industrial use.

Titanium dioxide is one of the popular industrial or engineering materials in our everyday life, e.g., titanium dioxide has widely been used as a white pigment for paints, cosmetics and foodstuffs. The photoactivity of titanium dioxide has been well-known for the past 60 years or more by reason of the phenomenon such as the chalking of white paints containing titanium dioxide after long-term outdoor exposure [36]. In the chalking, the organic components of the paints are decomposed as a result of the photocatalytic action of the titanium dioxide. Investigations on the active application of the photocatalysis of titanium dioxide were begun in 1950s [44]. The beginning of practically applicable photocatalysis of titanium dioxide originates with a discovery by Fujishima and Honda in 1972 [15] of photocatalytic splitting of water on titanium dioxide electrodes.

Titanium dioxide has three crystal structures, i.e., anatase, rutile and brookite. Titanium dioxide of the anatase type generally shows the highest photoactivity compared

Y. Ohama (✉)
Nihon University, Tokyo, Japan
e-mail: ohama@bh.wakwak.com

D. Van Gemert
Katholieke Universiteit Leuven, Heverlee, Belgium
e-mail: dionys.vangemert@bwk.kuleuven.be

Y. Ohama and D. Van Gemert (eds.), *Application of Titanium Dioxide Photocatalysis to Construction Materials*, RILEM State-of-the-Art Reports 5,
DOI 10.1007/978-94-007-1297-3_1, 

to the other types of titanium dioxide. Therefore, the anatase-type titanium dioxide is industrially utilized as an effective photocatalyst. The titanium dioxide photocatalyst ($TiO_2$ photocatalyst) is activated by solar light or ultraviolet radiation, whereby the following two phenomena and their effects are simultaneously induced on its surfaces by photocatalysis as seen in Fig. 1.2 [16, p. 154; 39, 47]:

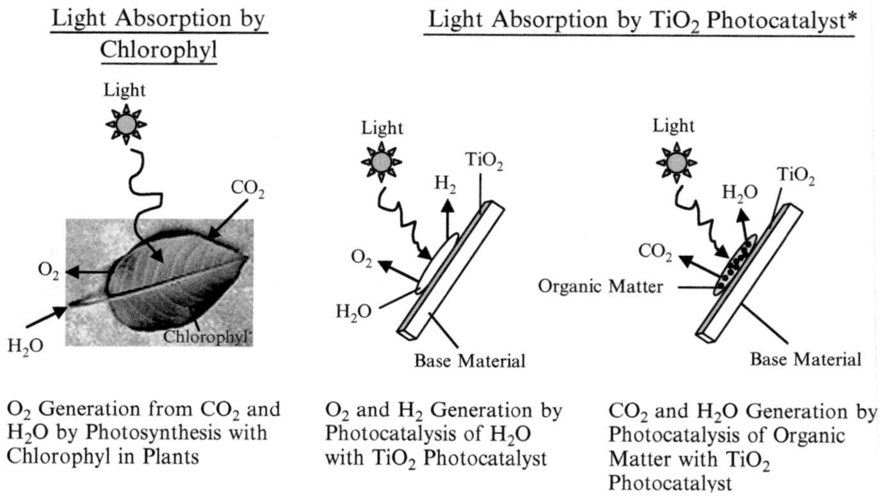


**Fig. 1.1** Comparison between $TiO_2$ photocatalysis and photosynthesis in plants. Note *: When there are $H_2O$ and organic matter on $TiO_2$ photocatalyst, the right photocatalysis precedes because the $TiO_2$ photocatalysis of the organic matter is faster than that of $H_2O$

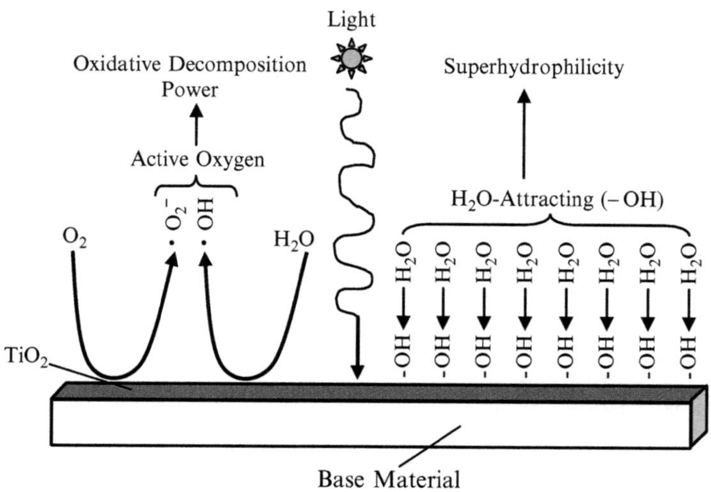


**Fig. 1.2** Two phenomena simultaneously induced by $TiO_2$ photocatalysis

1. Photocatalytic oxidative decomposition power due to active oxygen consisting of hydroxyl radicals (·OH) and superoxide ions ( $\cdot O_2^-$) – Strong decomposition power effect
2. Photoinduced superhydrophilication due to water-attracting hydroxyl groups (–OH) – High water-wettability effect

Such photocatalysis can also be induced by slight ultraviolet radiation due to fluorescent lights, etc. In 2001, the nitrogen-doped $TiO_2$ photocatalyst to enable a photocatalysis under visible light was developed [2], and has attracted much attention since then. More recently C-doped photocatalysts have been developed [6]. Construction materials with $TiO_2$ photocatalyst develop the major performance or functions as shown in Fig. 1.3 [16, p. 21; 24, p. 12], based on both effects of strong decomposition power and high water-wettability.

For the past 20 years, the technology of $TiO_2$ photocatalyst or photocatalysis has progressed rapidly, and become very attractive in the development of construction materials like ceramic tiles, blocks, glass, paints… with self-cleaning, air or water purification and antibacterial functions. As a result, various construction materials are being developed by utilizing the $TiO_2$ photocatalysis. A broad list of possible applications of photocatalytic cementitious materials can be summarized as follows [11]:

Horizontal applications

- concrete pavements
- paving blocks and paving plates
- coating systems for pavements and roads (whitetoppings, self-levelling mortars, …)
- roofing tiles
- roofing panels
- cement-based tiles

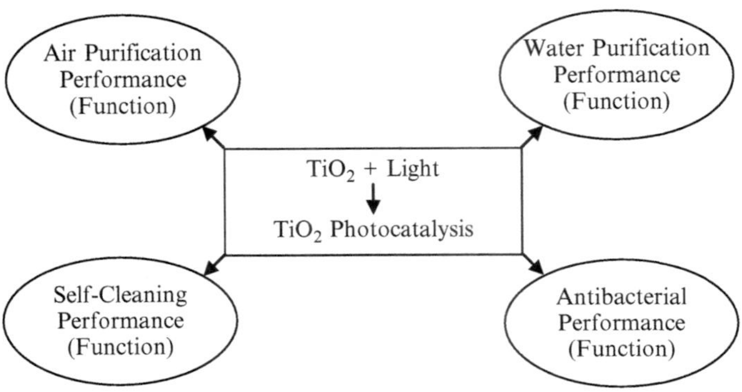


**Fig. 1.3** Major performance or functions developed by $TiO_2$ photocatalysis in construction materials with $TiO_2$ photocatalyst

Vertical applications

- indoor and outdoor paints
- finishing coatings, plasters and other final rendering cement-based materials
- permanent formworks
- masonry blocks
- sound-absorbing elements for buildings and roof applications
- traffic divider elements
- street furniture
- retaining fair-faced elements

Tunnels

- paints and renderings
- concrete panels
- concrete pavements
- ultrathin whitetoppings.

Life cycle analysis, examining and measuring the environment effects associated with the production of $TiO_2$ photocatalyst from the extraction of primary raw materials, through manufacturing, product use and eventual disposal, is a very complex issue. In limiting the study to $TiO_2$ manufacture, i.e., from cradle to gate, the overall environmental burden associated with different raw materials, processes, plants and waste treatment configurations can be determined. More data are being collected to evaluate the environmental effects of installed photocatalytic materials [7].

The present report reviews the history of the $TiO_2$ photocatalyst or photocatalysis, the principles of the $TiO_2$ photocatalysis, the development of construction materials with the $TiO_2$ photocatalyst, and the testing methods for construction materials with the $TiO_2$ photocatalyst.

# Chapter 2
# Principles of $TiO_2$ Photocatalysis

**Marta Castellote and Nicklas Bengtsson**

## 2.1 Definition of Photocatalysis

A substance can be thought to be a *catalyst* when it accelerates a chemical reaction without being consumed as a reactant; that is to say, it appears in the rate expression describing a thermal reaction without appearing in the stoichiometric equation [49]. A catalyst is a compound that lowers the free activation enthalpy of the reaction. Then, *photocatalysis* can be defined as the acceleration of a photoreaction by the presence of a catalyst [38, pp. 362–375]. This definition, as pointed out in [29, pp. 1–8], includes *photosensitization*, a process by which a photochemical alteration occurs in one molecular entity as a result of initial absorption of radiation by another molecular entity called the photosensitizer [13], but it excludes the photo-acceleration of a stoichiometric thermal reaction irrespective of whether it occurs in homogeneous solution or at the surface of an illuminated electrode. Otherwise, any photoreaction would be catalytic [29, pp. 1–8]. Depending on the specific photoreaction, the catalyst may accelerate the photoreaction by interaction with the substrate in its ground or excited state and/or with a primary photoproduct.

When the light is adsorbed by the catalyst C, the system represents a sensitised photoreaction which may occur through two different ways:

1. via energy transfer, by forming an activated state of the reactant of interest, S, which is more easily oxidized than their ground state:

$$C \xrightarrow{h\nu} {}^{*}C$$

$${}^{*}C + S \rightarrow {}^{*}S + C$$

$${}^{*}S \rightarrow P$$

M. Castellote (✉)
IETcc-CSIC, Spain
e-mail: martaca@ietcc.csic.es

N. Bengtsson
PhD researcher, IETcc-CSIC, Spain

Y. Ohama and D. Van Gemert (eds.), *Application of Titanium Dioxide Photocatalysis to Construction Materials*, RILEM State-of-the-Art Reports 5,
DOI 10.1007/978-94-007-1297-3_2, © RILEM 2011

2. via electron transfer, by acting either as an electron donor or acceptor.

$$C \xrightarrow{h\nu} {}^*C$$

$$^*C + S \rightarrow S^- + C^+$$

$$S^- \rightarrow P^-$$

$$P^- + C^+ \rightarrow P + C$$

In the case of energy transfer, the product P is formed from the activated substrate along the potential energy curve, while a new reaction path is opened when the photosensitizer transfers an electron to the substrate [29, pp. 1–8]. In considering excited-state redox reactivity, particularly that involved in organic photocatalysis, the direct occurrence of electron transfer will be of greater concern than involving energy transfer.

Due to the fact that the difference between a sensitised and a catalysed photoreaction is somewhat arbitrary, due to the different and complex mechanisms involved (static and dynamic sensitisation, interaction with a photoproduct, photoinduced reactions,…), the term, photocatalysis has therefore been defined as broad as possible without the specific implication of any special mechanism, and refers then to the action of a substance whose function is activated by the absorption of a photon. A photocatalyst can be described as one involved in the quantum yield expression for a photochemical reaction without its stoichiometric involvement; or more precisely, it appears in the quantum yield expression for reaction from a particular excited state to a power greater than its coefficient in the stoichiometric equation [14].

## 2.2 Oxidation of Organic Molecules by Photocatalysis

When the light of the appropriate energy illuminates the sensitizer, an electron from the valence band promotes to the conduction band, leaving an electron deficiency or hole, $h^+$, in the valence band and an excess of negative charge in the conduction band, $e^-$, which are oxidizing and reducing equivalents respectively and can participate in redox reactions.

The most important group of reactions photocatalysed by $TiO_2$ in construction materials are the oxidation of organic molecules. The redox potential needed to oxidize an organic substrate is determined by the position of the valence band and the redox potential of the organic substrate with respect to a standard electrode. If the organic substrate has a more negative redox potential than the redox level of the photogenerated hole, it may reduce $h^+$ giving the cation radical of the organic substrate, $S^{\bullet+}$. Subsequent reaction of $S^{\bullet+}$, that is faster than back electron transfer, leads to product formation. On the other hand, if the hole is reduced by water or adsorbed $OH^-$ ions, HO• and/or other radicals are formed, able to oxidize organic matter. The electron is taken up by an oxidizing agent, normally the adsorbed oxygen.

In environmental chemistry, from a functional point of view, the photocatalysis is one of the so-called Advanced Oxidation Technologies (AOT's). The concept of AOT's was initially established by Glaze and coworkers [20], who defined the

AOT's as the process involving the generation and use of powerful redox transient species, mainly the hydroxyl radical (HO·). This radical can be generated by different procedures, and is very effective in the oxidation of different species, mainly organic matter.

Among the nonphotochemical AOT's, one can distinguish the oxidation with $O_3/OH^-$, $O_3/H_2O_2$, Fenton's processes, electrochemical oxidation, radiolysis, plasma, ultrasonic treatment, etc. Among the photochemical processes, we can find the oxidation in subcritical and supercritical water, photolysis of water in UVV, $UV/H_2O_2$, $UV/O_3$, $UV/H_2O_2/O_3$, Photo-Fenton's processes, and heterogeneous photocatalysis.

The hydroxyl radical, OH·, exhibits the main characteristic of being able to attack to every organic compound; after the fluorine, it is the most powerful oxidizing agent and reacts $10^6$–$10^{12}$ times faster than alternative agents, as $O_3$ [8]. That gives an idea of the oxidation potential of the photocatalysis.

## 2.3 Other Chemical Concepts of Interest

In the field of heterogeneous photocatalysis, there are several fields of the chemistry that are of great importance. Among them, the solid state chemistry [22, 35, 45], with the knowledge on crystal structure, nonstoichiometry and defect chemistry, electronic properties of solids, chemical bonds, structures and arrangements of atoms, electronic conductivity, are remarkable. In addition, the knowledge of surface chemistry is very important, especially the structures, compositions, dynamics of atoms and molecules, and the electronic properties of surfaces. The adsorption and bonding of gas atoms and molecules on surfaces are also needed for understanding the mechanisms in heterogeneous photocatalysis [1, 28, 46].

## 2.4 Principles of $TiO_2$ Photocatalysis

Titanium dioxide ($TiO_2$) is an excellent photocatalyst with applications in various fields. The main advantages of $TiO_2$ are its high chemical stability when exposed to acidic and basic compounds, its nontoxicity, its relatively low cost and its highly oxidizing power, which make it a competitive candidate for many photocatalytical applications [5, 9, 16, 17, 41]. $TiO_2$ exists in three different crystalline modifications: anatase, brookite and rutile, where anatase exhibits the highest overall photocatalytic activity. Basic physical properties of the anatase-type $TiO_2$ are listed in Table 2.1 [16, p. 125].

**Table 2.1** Basic physical properties of anatase-type titanium dioxide

| | |
|---|---|
| Crystal form | Tetragonal system |
| Density ($g/cm^3$) | 3.90 |
| Refractive index | 2.52 |
| Permittivity | 31 |
| Thermal stability | Change to rutile type at high temperature |

The highly oxidizing effect of $TiO_2$ makes it suitable for decomposition of organic and inorganic compounds at very low concentrations ranging from 0.01 to 10 ppm [16]. The photocatalytic effect of $TiO_2$ can be used for self-cleaning surfaces, decomposing atmospheric pollution and self-sterilization [9, 17]

### 2.4.1 Oxidation-Reduction Reactions

$TiO_2$ is a semiconductive material that during illumination acts as a strong oxidizing agent lowering the activation energy for the decomposition of organic and inorganic compounds. The illumination of the surface of the $TiO_2$ induces the separation of two types of carriers: (1) an electron ($e^-$) and (2) a hole ($h^+$). To produce these two carriers, sufficient energy must be supplied by a photon to promote an electron ($e^-$) from the valence band to the conduction band, leaving a hole ($h^+$) behind in the valence band. The recombination of holes and electrons is relatively slow in $TiO_2$ compared to electrically conducting materials, i.e., metals where the recombination occurs immediately.

$$TiO_2 + hv \rightarrow h^+ + e^-$$

The required energy that has to be supplied by the photons for the promotion of the electrons depends on the band gap for the specific material. The band gap is the difference in energy between the highest permitted energy level for the electron in the valence band and the lowest permitted energy level in the conduction band. The band gap is the minimum energy of light required to make the material electrically conductive [5]. The band gap energy, $E_g$ of $TiO_2$ (anatase) is 3.2 eV, which corresponds to photons with a wave length of 388 nm [16].

The photoinduced hole can oxidize a donor molecule (D) adsorbed on the $TiO_2$ surface.

$$D + h^+ \rightarrow \cdot D^+$$

The electron in the conduction band can reduce an acceptor molecule (A).

$$A + e^- \rightarrow \cdot A^-$$

The strong oxidation power of the hole enables a one-electron oxidation step with water to produce a hydroxyl radical ($\cdot OH$).

$$H_2O + h^+ \rightarrow \cdot OH + H^+$$

Oxygen can act as an electron acceptor, and be reduced by the promoted electron in the conduction band to form a superoxide ion ($\cdot O_2^-$). The superoxide ion is a highly reactive particle, able to oxidize organic materials. The oxidation-reduction process is shown in Fig. 2.1.

$$O_2 + e^- \rightarrow \cdot O_2^-$$

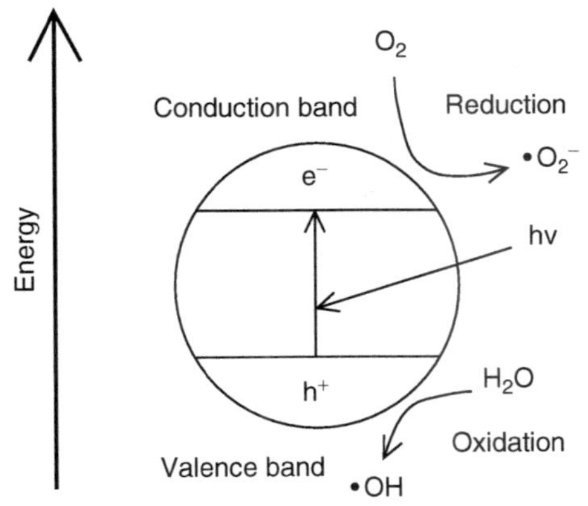


**Fig. 2.1** Photopromotion of an electron, reduction of oxygen and oxidation of water

## *2.4.2 Development of Superhydrophilicity*

The degree of water repellence on a surface of a specific material can be measured by the water contact angle. A hydrophobic surface has a water contact angle that is higher than a hydrophilic surface. For glass and other inorganic materials, the water contact angle is between 20° and 30°. Almost no surface is known with a water contact angle less than 10°, compared to the water contact angle of the illuminated surface of titanium dioxide being less than 1° [16].

The fogging effect that can be experienced on the bathroom mirrors and the windows of cars is caused by the condensation of water damp that forms small droplets on the surfaces. Until now, the main approach to prevent this type of fogging has been to create a hydrophobic surface that repels the water molecules. This method still leads to creation of water drops that have to be removed from the surfaces for example by blowing or shaking.

The superhydrophilic effect of $TiO_2$ is formed when the surface is exposed to UV light, and after a certain time of moderate illumination the water contact angle approaches zero [17]. When the illumination stops, the superhydrophilic effect disappears. If the surface is prepared with a water-retaining material like silicone dioxide or silica gel, the superhydrophilic effect can be maintained even after the light is turned off [16]. On the surface of a hydrophilic material, the condensation of water damp forms a uniform film, which flattens out instead of fogging the surface with water drops.

The superhydrophilic effect is also caused by to the production of holes because the electrons tend to reduce the Ti(IV)-cations to Ti(III)-ions

$$e^- + Ti^{4+} \rightarrow Ti^{3+}$$

and the holes oxidize the $O_2^-$ -anions.

$$4h^+ + 4O_2^- \rightarrow 2O_2$$

This process leads to ejection of oxygen atoms and creation of oxygen vacancies at the $TiO_2$ surface. These vacancies are covered by water molecules that are forming OH-groups that create the superhydrophilic effect.

# Chapter 3
# Application of $TiO_2$ Photocatalysis to Cementitious Materials for Self-Cleaning Purposes

**Anibal Maury Ramirez and Nele De Belie**

## 3.1 Introduction

Cementitious materials, especially those exposed to outdoor conditions, are directly and continuously exposed to many atmospheric pollutants, (organic, inorganic and particulate matter), microorganisms (e.g., algae, fungi, cyanobacteria) and different weather conditions. As a result, they have an accelerated deterioration process which can produce in many cases important changes in the materials properties, among them aesthetic properties. Colour, for example, has to give a pleasant appearance that should give to the public an adequate perception of the quality and maintenance of buildings or structures. However, serious colour changes are produced on the buildings and structures due to the mineral composite nature of cementitious materials. The nature causes relatively high porosity and roughness, which enable the deposition of coloured organic pollutants or particulate matter and partially facilitate biological growth.

In order to avoid and control colour changes of cementitious materials, common practices such as the use of additives, sealers, chemical cleaners, paintings, restoration and repairing works are widely used in buildings or structures. Nevertheless, this effort, though valuable, has not led to a complete elimination of the colour changes because of the complexities and variability of natural environments, the multi-component nature of cementitious materials and finally the effects of atmospheric pollutants and micro-organisms on cementitious materials [18, 32]. Additionally, apart from the direct costs of these activities, indirect costs related with operational and environmental losses should be considered. It is therefore important to tackle these problems by formulating efficient, innovative and environmentally friendly cementitious materials.

A.M. Ramirez
PhD researcher, Ghent University, Belgium
e-mail: Anibal.MauryRamirez@UGent.be

N. De Belie (✉)
Ghent University, Ghent, Belgium
e-mail: Nele.DeBelie@UGent.be

Y. Ohama and D. Van Gemert (eds.), *Application of Titanium Dioxide Photocatalysis to Construction Materials*, RILEM State-of-the-Art Reports 5,
DOI 10.1007/978-94-007-1297-3_3, 

Application of titanium dioxide photocatalysis on cement-based materials has enabled the degradation of a range of organic compounds (e.g., VOC) and some inorganic compounds ($NO_x$ and $SO_2$) which are aggressive towards cementitious materials properties and the environment. Thus, $TiO_2$ could be used to increase the life cycle of cement-based materials while it could also substantially decrease the concentration of some air pollutants in urban air, especially in confined spaces such as canyon streets, tunnels and parking lots or very polluted open spaces such as gas stations or some specific industries.

Since the early stage of the twentieth century, rutile-type titanium dioxide was used as a pigment for paintings, cosmetics and others. However, initially the high ability of $TiO_2$ to degrade organic compounds destroyed the organic binders, producing the painting phenomena known as "chalk effect". Today, for avoiding the degradation of organic binders, $TiO_2$ pigments are very often coated with different oxides, like alumina, silica and zirconia which reduce the amount of produced radicals and oxides. Only since 1969, the research work of Fujishima and Honda on the photocatalytic splitting of water using titanium dioxide electrodes (anatase type) made it possible to use the photocatalytic properties with other goals [15].

Initially, the antibacterial properties of titanium dioxide were used in air and water purification systems [19, 25, 43]. Then, ceramic tiles with titanium dioxide coatings were used for creating antiseptic effect in rooms from hospitals, care facilities, schools, kitchens and baths [5]. Today, addition of metals such as Cu and Ag to $TiO_2$ has increased the antibacterial properties [6]. One important accidental finding happened in 1995: $TiO_2$ films, prepared with a certain percentage of silica ($SiO_2$), developed a hydrophilic behaviour when they were exposed to UV light [48]. According to different experiments, hydrophilicity appears to be related to the density in surface $^{\bullet}OH$ groups. This density increases under UV irradiation. Water molecules are prone to be hydrogen–bound to the $^{\bullet}OH$ groups, which explain why rainwater forms films instead of depositing as droplets [17, 48]. In general, when there is no UV light and $TiO_2$ is used alone, the hydrophilic behaviour is stopped. However, when $TiO_2$ is combined with silica or a chemical compound of silicon that has siloxane bonding, the hydrophilic effect of the surface continues for several days even in the dark [42]. Nowadays, the hydrophilic behaviour of $TiO_2$ has been extensively applied commercially in glasses and mirrors for keeping the optical properties of these materials under weather conditions that are favourable for the water vapour condensation. Recently, a new approach to obtain self-cleaning surfaces has been announced. Introducing roughness to $TiO_2$ particles by a photoelectrochemical etching technique could cause a controlled hydrophilic behaviour and a hydrophobic behaviour in the same surface [23]. This discovery will certainly expand the application field of $TiO_2$ for self-cleaning purposes.

However, the hydrophilic/hydrophobic behaviours produced by $TiO_2$ photocatalysis are more pronounced on smooth surfaces like glass and ceramic tiles than in cementitious materials surfaces. In the latter materials, this effect is limited due to their relatively high roughness, porosity and permeability. Nevertheless, due to $TiO_2$ photocatalytic action, photodegradation of organic staining substances on cementitious materials could take place, allowing later an easy cleaning process with water (e.g., with rain water). In the case of inorganic powders, because these adhere to the surfaces of cementitious materials through an organic-based interlayer, degrading this layer by $TiO_2$ photocatalysis, again a later cleaning process using water will be easier.

## 3.2 Self-Cleaning on Cementitious Materials

Considering the successful results obtained in the previous applications of $TiO_2$ and, on the other hand, the conditions offered by building materials from cities (sunlight exposure and considerable surface areas), titanium dioxide photocatalysis has been applied in different cementitious materials. In this field, the applications have varied from self-cleaning to air-cleaning purposes. However, the first applications were intended to obtain an automatic cleaning process on buildings facades, especially for materials based on white cement. Since then, the self-cleaning activity of cementitious materials loaded with $TiO_2$ has been evaluated through laboratory tests using organic dyes as pollutant models. In most of the cases on laboratory scale, the photocatalytic activity has been initiated using different types of lamps with UV rays.

In 1996, the first laboratory experiments on the self-cleaning activity of titanium dioxide contained in cementitious materials for designing the mix compositions of the church, "Dives in misericordia" (Rome, Italy) and for evaluating the self-cleaning ability of a new type of cement (TX Millennium) were reported [10, 27].

The promising results of the titanium dioxide self-cleaning activity obtained on laboratory scale allowed researchers to continue to another stage for developing self-cleaning cementitious materials. Hence, some buildings with $TiO_2$-loaded materials have been built in Europe and Asia.

The results from the pioneering European buildings using $TiO_2$, the Church "Dives in Misericordia" and the Music and Arts City Hall, are promising, and a tendency to keep original colour has been evidenced [21]. However, for an appropriate evaluation of the photocatalytic action of $TiO_2$ in these applications, the correlation between the colour changes produced on the surfaces, the environmental conditions and the architectural plans should be further investigated for longer times. Besides, the comparison between the colour changes produced on the building surfaces and some reference panels or samples could be advisable.

In relation to microbial growth on cementitious materials, mainly due to the high roughness, high porosity and rich mineral compositions of cementitious materials, the generation of different microorganisms such as algae, fungi and cyanobacteria is a common phenomenon. The cyanobacteria has the potential to cause surface staining, produces chemical changes and decreases the durability of cement-based materials from buildings and structures [30].

Titanium dioxide has shown an ability to decompose organic compounds from the outer cell membranes of some microorganisms, which cause cell death. Based on this, $TiO_2$ has been studied to eliminate *E. coli* in water treatments [43]. Also, using titanium dioxide (Degussa P-25) complete photodegradation of algae

(*Cladophora*) has been reported. In vitro experiments showed that *Cladophora* samples were physically and biologically degraded when subjected to $TiO_2$ mediated photocatalysis. For the most successful photocatalytic process, $TiO_2$ was immobilized on a glass surface, and used in combination with either sunlight or artificial UV light. The loss of vital algae pigments was monitored using UV-vis spectrophotometry, and cell structural changes were determined by microscopy observation. *Cladophora*, in the presence of $TiO_2$ covered glass beads, experienced a loss of chloroplast pigments after 2 h of UV lamp light irradiation (450 W Hg vapor lamp). In a separate experiment, sunlight exposure over 4 days (~24 h) resulted in the complete oxidative degradation of the green chloroplast pigments, verified by the UV spectra of the algal extracts [37]. These results suggest that $TiO_2$, immobilized on cementitious materials may be useful in controlling algae growth, thus mitigating many of the economic, aesthetic and ecological impacts of algae growth on cementitious materials.

Using a recently developed equipment for accelerated algae growth on cementitious materials, preliminary tests for evaluating algicidal effect of $TiO_2$ coatings applied on autoclaved aerated concrete samples were performed during 5 weeks in the Magnel Laboratory (Ghent University, Belgium). The equipment consists of 12 independent compartments for different sample types. One single compartment is composed of inclined stainless steel surfaces to place the samples, an aerated reactor for growing an algae suspension and a pumping system to periodically scatter the algae suspension on the samples surfaces during 90 min. Daytime and nighttime period were reproduced every 12 h. Daytime periods were simulated with Grolux lamps (30 W). The illumination amounted to $800 \pm 84$, $524 \pm 62$ and $448 \pm 114$ $l_x$ at the upper and lower parts of the stainless steel compartment. After each test week, the evaluation of the algae growth was done by image analyses and colorimetric measurements. Colorimetric measurements were performed by using an X-Rite SP60 colorimeter. Three measurements were taken at standard positions on the samples. For the image analysis, the photographs of the samples were obtained by the use of a Canon Scan 3000 F scanner. The images were processed by means of image 1.38x software [12].

As algae growth is a biological process, two previous weeks were needed to obtain the optimal growing conditions of the algae in the set-up, so the results from this period are not representative and as a consequence not presented here. Results from image analysis can be seen in Fig. 3.1. In relation to reference concrete (uncoated samples), $TiO_2$-coated concrete evidenced the partial photodegradation of *Chlorella vulgaris*. During the first 2 weeks, the reference concrete and the $TiO_2$-coated concrete behaved similarly, showing no significant growth of algae on their surfaces. However, after the third week, a substantial change was evidenced, and an increase in the algae growth was observed in both concretes. Specially the reference concrete showed an area covered with algae, equal to 75%, while the $TiO_2$-coated concrete gave an area of only 25%. After the fourth week, an increase in the area covered with algae was noted again on both concretes. In particular, the reference concrete was covered at an area of almost 100%, while $TiO_2$-coated concrete was covered at an area of only 60%. After the whole test (5 weeks), $TiO_2$-coated concrete presented a constant covered algae-area of around 60%. This value is promising

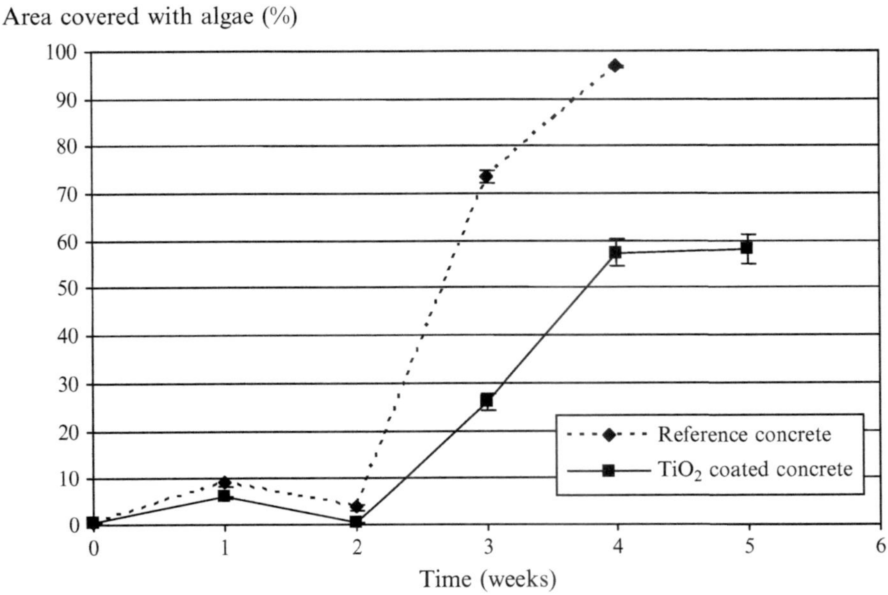


**Fig. 3.1** Area covered with algae on autoclaved aerated concretes after accelerated algae growth test for 5 weeks

because of the accelerated conditions for the algae growth during the test, and also because the algae photodegradation was obtained with UV light coming from fluorescent light (30 W) which was even periodically switched off. Furthermore, weathering resistance problems on the $TiO_2$ coatings were evidenced during the test, so that the $TiO_2$-coated concrete did not show its full algicidal capacity during the complete test.

Even when the pioneering European buildings using $TiO_2$ evidenced a good photocatalytic activity for degrading colored pollutants from air, major efforts to obtain $TiO_2$ coatings for cement-based materials should be done for making this technology more popular. Also, for the application of $TiO_2$ as an additive in the mixes, more studies should be done in relation to an apparent increase in the frost resistance and the negative effect of $TiO_2$ additions on organic additives.

Considering the magnitude of the microbial growth on external cementitious materials and the known bactericidal and algicidal effect of $TiO_2$ on other materials, it was a surprising fact that studies about these properties on $TiO_2$-loaded cementitious materials were not found. The recent results obtained by Magnel Laboratory for Concrete Research (Ghent University) about the use of titanium dioxide (anatase-type) coatings for algaecide coatings on autoclaved aerated concretes are promising.

The application of $TiO_2$ photocatalysis to cementitious materials for self-cleaning purposes is a topic which still leaves many questions unanswered and has the reason to continue researching different aspects. In a few years, this novel technology seems to become a topic of popular interest due to the impact of the self-cleaning and air-purifying properties of $TiO_2$ on the life quality of the future cities [9].

# Chapter 4
# Application of Antibacterial and Self-Cleaning Effects to Noncementitious Construction Materials

**Yoshimitsu Saeki**

## 4.1 Introduction

Recently, photocatalyst coating technology enabled to produce construction materials having advanced functions such as sterilizing, deodorizing and self-cleaning properties.

In 1995, the novel phenomenon of the photocatalyst was discovered in Japan. That is, when the surface of photocatalyst is exposed to light, the contact angle of the photocatalyst surface with water is reduced gradually. After enough exposure to light, the surface reaches superhydrophilicity. In other words, it does not repel water at all, so the water cannot exist in the shape of a drop, but spreads flatly on the surface of the photocatalyst.

One of the most popular applications of photocatalyst is for its self-cleaning capability in exterior construction materials of buildings. There are various exterior finishing materials treated with photocatalyst on the Japanese market at present.

Surface appearance of a building is one of the esthetic visual user requirements, which is further composed of several subfactors such as color, texture, regularity, flatness and so on. As to the soiling of exterior construction materials, it can be divided into the uniform one and the local one, which stands for uniformly soiled and partial or dotted soiling within a certain area of a building plane or a component. In general, the latter is much noticeable than the former one. Various countermeasures to avoid soiling had been tried both from design aspects such as shape, dimension, details of external component as well as from material modification aspects by changing water-repellent property, hydrophilic property, electrostatic property and so on. Application of photocatalyst is one of the new countermeasures.

Y. Saeki (✉)
Toto Ltd., Japan
e-mail: yoshimitsu.saeki@toto.co.jp

Y. Ohama and D. Van Gemert (eds.), *Application of Titanium Dioxide Photocatalysis to Construction Materials*, RILEM State-of-the-Art Reports 5,
DOI 10.1007/978-94-007-1297-3_4, © RILEM 2011

## 4.2 How to Form Photocatalyst Films on Surfaces of Construction Materials

To apply these photocatalytic properties of $TiO_2$ to the products, several different techniques are needed in order to form $TiO_2$ thin films on the substrates. These techniques may vary considerably depending on the substrates, for example, tile, glass or plastics. Therefore, photocatalytic coating agents applicable to various materials are needed, and indeed are being developed.

These coating agents may be divided into two categories. The first category consists of coating agents that require high-temperature heat treatment to harden the titanium dioxide films, based on the existing thin-film-forming technique. The second category consists of coating agents that make use of a low-temperature curing step made possible by the compounding of polymers and inorganic compounds.

At first, high-temperature heat treatment technique has been developed. That is, for the glazed tile or ceramic substrates, a mixture dispersion of $TiO_2$ and $SiO_2$ sol is coated on the surfaces, and then heated to 600–800°C. With heat treatment, the $TiO_2$ layers are sintered, and strongly attached to the surfaces. Figure 4.1 illustrates the process for forming photocatalytic thin film on tile [42]. It is characterized by the fact that the material is sintered for a short time for fixation after it is spray-coated with photocatalyst dispersion to 10 nm fine particles of $TiO_2$. The thickness of the photocatalytic thin film is controlled at around 150 nm.

As the high-temperature heat treatment techniques for making $TiO_2$ coatings are limited to heat-resistant substrates, low-temperature-curing type coating agents have also been developed. These are coating agents that can be applied to general use plastics at curing temperatures ranging from 60°C to 120°C. With this technology, a double-layer coating is used, so that the photocatalyst layer adheres to the substrate plastics while protecting the plastics from the harsh effects of photocatalytic oxidative decomposition. Although the coating and curing steps must be performed twice with this technique, the benefits are expected to outweigh the additional processing. It is applicable to almost all types of substrate materials such as plastics, paints, metals, glass and tile.

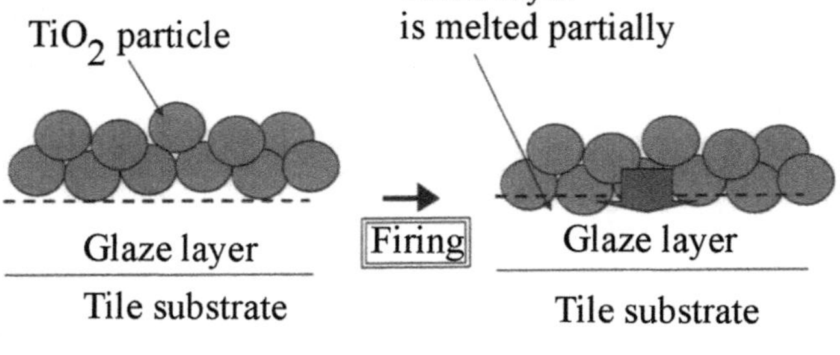


**Fig. 4.1** Process for forming $TiO_2$ thin film on tile

## 4.3 Functional Design and Supplement of Photocatalytic Effect in the Dark

Supplement of the characteristics of the photocatalytic reaction in the dark is highly important to apply photocatalyst in practical use. Commercialized $TiO_2$ coating demonstrates hydrophilic and decomposition features at varying rates depending on the composition and structure. Different combinations of the two characteristics are suited to respective applications.

### *4.3.1 Addition of Silica- or Silicon-Based Water-Retaining Substances*

With $TiO_2$ photocatalyst alone, the superhydrophilicity disappears shortly once the exposure of the surface to UV light is stopped. When the photocatalyst is combined with silica- or silicon-based chemical compounds that have siloxane bonding, the hydrophilic effect of the surface continues for a long time even in the dark, as seen in Fig. 4.2 [42].

### *4.3.2 Addition of Antibacterial Metals Such as Cu and Ag*

Antibacterial, deodorizing and self-cleaning functions of photocatalyst derive from the organic compound decomposition ability of active oxygen, generated from photocatalytic reaction. For the purpose of improving oxidative degradation activity, it is essential to reduce the recombination of photo-excited electron-hole pair.

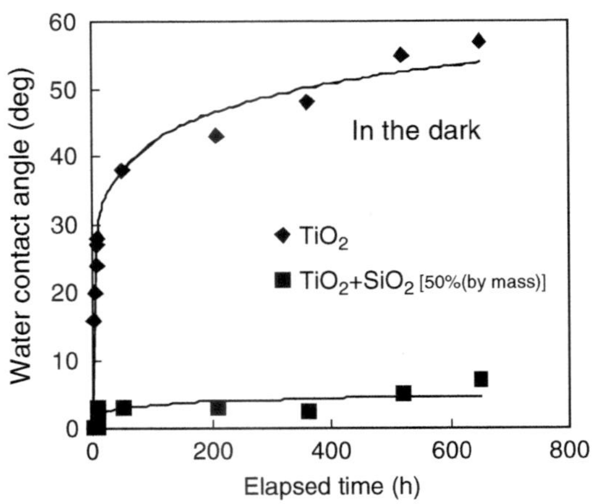


**Fig. 4.2** Effect of silica addition to sustain hydrophilicity

For applications focusing on the sterilizing feature, photocatalysts are normally incorporated with antibacterial metals such as Cu and Ag to exert the antibacterial function even in the dark. Metal incorporation also improves the decomposition activity when the photocatalyst is exposed to light because it produces a more efficient charge separation of the electron and electron hole generated from photoexcitation.

## 4.4 Photocatalysts Applied to Noncementitious Construction Materials

### *4.4.1 Exterior Construction Materials*

The advantages of the photocatalyst applied in exterior materials lie in esthetic preservation, reduction of life cycle costs (LCC) including cleaning cost. Figure 4.3 shows a comparison of the self-cleaning properties of exterior tiles actually installed. Since their release in October 1998, photocatalyst tiles have proved so successful that they have been introduced to more than 5,000 projects to date.

Photocatalyst is also expected to work as a solution to inhibit the mold and algae that plague outdoor structures. Figure 4.4 shows a comparison of the anti-algae properties of photocatalyst paint. Usual antialgae paint has no antialgae property left after 3 years in an accelerated test due to the dissolving out of anti-algae agent. However, photocatalyst paint keeps sufficient property even after 12 years of accelerated test.

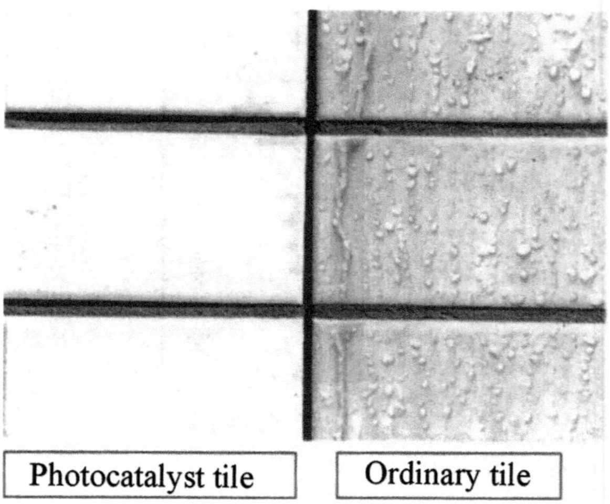


**Fig. 4.3** Comparison in self-cleaning property of photocatalytic exterior tile and ordinary tile

**Fig. 4.4** Comparison in antialgae property of photocatalyst paint

### *4.4.2 Interior Construction Materials*

As for interior materials, photocatalyst technologies are increasingly used with tiles, paints and window shades. In addition to the conventional sterilizing function, the easy-to-clean feature based on hydrophilicity makes these products less likely to be stained with greasy dirt in the kitchens or with soap scum in the bathrooms and also makes it easier to wash the dirt off with water. They have a wide spectrum of applications, ranging from public lavatories, restaurant kitchens and clinical operation rooms to kitchens, bathrooms and washrooms at home.

Applications of photocatalyst have been spreading widely in Japan as shown in Fig. 4.5. The gross sales of Japanese photocatalyst market in 2005 were about 60 billion yen and two-third of them concerned construction materials (researched by Photocatalysis Industry Association of Japan).

Photocatalyst technologies and applied products are expected to meet the demands of society, such as conserving the global environment as well as preserving the aesthetics of buildings and prolonging their lives. The need to establish unified assessment methods on functions and performance of photocatalysts becomes urgent.

To encourage the sound development of both photocatalytic technologies and their market, Photocatalysis Industry Association of Japan (PIAJ) runs campaigns for market awareness and sales expansion of photocatalyst products and, in technical terms, works on drafting standards for testing performance and normalization of photocatalytic products.

**Fig. 4.5** Photocatalytic construction materials in Japan

# Chapter 5
# Applications of $TiO_2$ Photocatalysis for Air Purification

**Anne Beeldens, Luigi Cassar, and Yoshihiko Murata**

## 5.1 The $TiO_2$-and-Cement System

A system consisting of $TiO_2$ and cement has recently been studied within the framework of a strategy for alleviating environmental pollution through the use of construction materials containing photocatalysts. This technology for the photocatalytic degradation of organic pollutants is aimed at maintaining the aesthetic characteristics of concrete structures, particularly those based on white cement. The main reason for the discoloration of cementitious materials is the accumulation of coloured organic compounds on their surfaces (see also Sect. 3.2). Suitable amounts of $TiO_2$ have been introduced into cement mixes to render the surfaces of the resulting structures photocatalytically active. In order to verify the photocatalysis, experiments focusing on the oxidation of several kinds of aromatic organic compounds have been carried out. For instance, white cement disks have been impregnated with a phenanthroquinone solution (0.1 mg/cm$^2$), yielding homogeneously yellow surfaces. Accelerated irradiation tests were then performed with a solar simulator (100 h of irradiation, corresponding to 1 year of sunlight). A rapid restoration of the clean surface was obtained. These results open the way to the widespread development of self-cleaning concrete and mortar-coated walls. $TiO_2$/cement composites are expected to maintain their aesthetic characteristics unchanged over time, in particular, the

A. Beeldens (✉)
Belgian Road Research Center, Belgium
and
Katholieke Universiteit Leuven, Heverlee, Belgium
e-mail: a.beeldens@brrc.be

L. Cassar
CTG Italcementi Group, Italy
e-mail: l.cassar@itcgr.net

Y. Murata
Ube-Mitsubishi Cement Research Institute, Japan
e-mail: murata@mmc.co.jp

Y. Ohama and D. Van Gemert (eds.), *Application of Titanium Dioxide Photocatalysis to Construction Materials*, RILEM State-of-the-Art Reports 5,
DOI 10.1007/978-94-007-1297-3_5, © RILEM 2011

**Fig. 5.1** Dives in Misericordia

colour, even in the presence of aggressive urban environments. A white cement containing $TiO_2$ has already been used for the construction of a school in Mortara, Italy (completed in 1999), the Cité de la Musique in Chambéry, France (completed in 2000), and the *Dives in Misericordia* Church in Rome (completed in 2003, see Fig. 5.1). Cementitious paints containing photocatalysts have also been developed, and applied on residential buildings in Italy. A particularly interesting aspect of $TiO_2$/cement composites is that there is a clear synergy between the cement and $TiO_2$ that makes cement an ideal substrate for environmental photocatalysis. Many photo-oxidizing compounds such as $NO_2$ and $SO_2$ are acidic. The basic nature of the cement matrix is particularly suitable for fixing both the polluting reagent and the photo-oxidation products at its surface. Another important aspect is the use of nanometric $TiO_2$, which increases substantially the photoactivity of the $TiO_2$/cement system.

## 5.2 Air Purification Systems by Removing $NO_x$

A variety of air pollutants have known or suspected harmful effects on human health and environment. In most areas of Europe, these pollutants are principally the products of combustion from space heating, power generation or from motor vehicle traffic. Pollutants from these sources may not only prove a problem in the immediate vicinity of these sources but can travel long distances, chemically reacting in the atmosphere to produce secondary pollutants such as acid rain and ozone.

The main pollutants emitted by vehicles are carbon monoxide, oxides of nitrogen ($NO_x$), volatile organic compounds (VOCs) and particulates. These pollutants have an increasing impact on the urban air quality. In addition, the photochemical reactions resulting from the action of sunlight on $NO_2$ and VOCs lead to the formation of ozone, a secondary long-range pollutant, which impacts in rural areas often far from the original emission site. Acid rain is another long-range pollutant influenced by vehicle $NO_x$ emissions and resulting from the transport of $NO_x$, oxidation in the air into $NO_3$ and finally precipitation of nitric acid with harmful consequences for building materials (corrosion of the surfaces) and vegetation.

$TiO_2$-loaded cementitious materials are interesting not only for their self-cleaning properties. Research under way shows that they also have good potential in urban pollution control. Examples of pollutants that can be eliminated by the photocatalytic $TiO_2$/cement system are $NO_x$, $SO_x$, $NH_3$, CO, volatile aromatic hydrocarbons such as benzene and toluene, organic chlorides, aldehydes, and polycondensated aromatics. The mechanism of nitrogen oxide removal by photocatalysis is not simple. It is assumed that NO in the air is oxidized when the catalyst is exposed to light. Through the intermediate step of nitrogen dioxide ($NO_2$), it is then converted to nitrate. When $NO_2$ is formed, part of the gas may escape from the photocatalytic surface, but in the presence of the cement matrix, the gas may be effectively trapped, together with the nitrate salt formed.

To determine the efficiency of the photocatalyst, different laboratory testing methods may be applied. Some of these are already standardized (see Chap. 6).

Another method (not standardised), which was used in the PICADA (Photocatalytic Innovative Coverings Applications for De-pollution Assessment) project [31, p. 69], consists of a testing chamber with a certain surface of photocatalytic materials at the walls of this chamber. The photoconversion of $NO_x$ is looked at over time. The percentage is referred to an initial concentration of 100%. In the case of cementitious materials as base materials, a concentration decrease is observed even without lumination, most probably due to adsorption by the hydroxides present in the cement matrix. However, the removal of $NO_x$ by the action of the photocatalyst/cementitious matrix composite is clearly much higher than the removal obtained with the photocatalyst alone or with any other combination, in the dark or in the light. Thus, once more, a synergistic effect is obtained.

A parameter evaluation in the laboratory revealed the influence of temperature, relative humidity and contact time (surface, flow velocity, height of the air flow over the sample,…). In general, it can be stated that the efficiency towards the reduction of $NO_x$ increases with a longer contact time (larger surface, lower velocity and higher turbulence), with a higher temperature and lower relative humidity. These are the conditions at which the risk of ozone formation is the largest: high temperatures, no wind and no rain. At these days, the photocatalytic reaction will be more present.

## 5.3 Photocatalytic Blocks

### 5.3.1 *Feature of Photocatalytic Blocks*

A photocatalytic block is an interlocking block or pavement (paving) block that contains titanium dioxide in its top surface. It is chiefly used for road pavements. The photocatalytic blocks can purify air using light energy. The atmospheric $NO_x$ pollution due to exhaust gases from automobiles has been a serious problem in urban areas. The photocatalytic block pavement is one of the measures for $NO_x$ pollution control. It was developed by Mitsubishi Materials Corporation [33] with the aim of purifying the environment using solar energy in Japan in 1996. The blocks have the surface layers of titanium dioxide contained in a cement mortar which is monolithically molded with the substrates by using a vibration press. The thickness of the surface mortar layers is about 7 mm. They are used in pavements (side-walks), roadways, parks, and building peripherals.

The photocatalytic blocks use cement in order to immobilize a photocatalyst. Cement has excellent features as a binder for the photocatalyst to purify the air. Figure 5.2 illustrates the $NO_x$ removal schematic of the photocatalytic block [4] The $NO_x$ removal of the photocatalyst is effective because the hardened body of the cement is porous. Moreover, the active oxygens generated by the light adsorption of the photocatalyst cannot oxidize the cement, which is already a mixture of oxides. The nitrate ions or nitric acid generated by photocatalysis are neutralized depending on alkaline dusts in the environmental air. The photocatalytic blocks have sufficient durability. The following advantages are suggested when the photocatalytic blocks are used for pavements.

1. The blocks can be used as flat outdoor surfaces, where sunlight and rain necessary for the activation and regeneration are available.

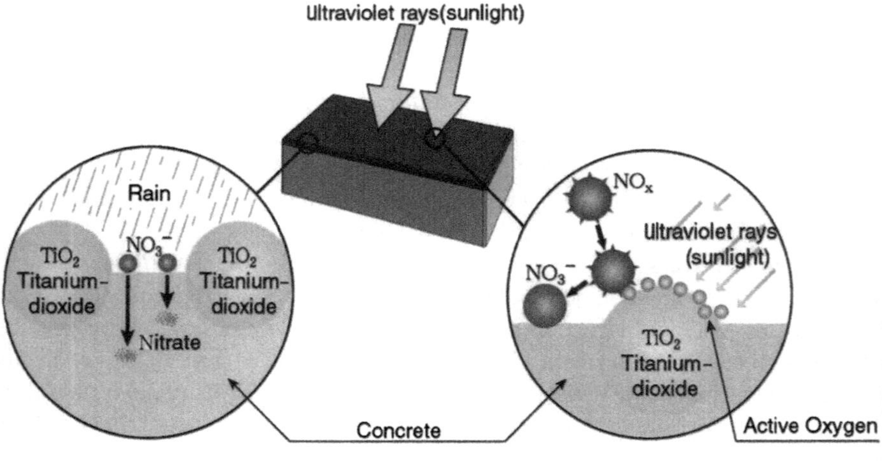


**Fig. 5.2** $NO_x$ removal schematic of photocatalytic block

2. The blocks can be installed close to the $NO_x$ emission sources, and react with pollutants under a higher concentration.
3. The blocks can be adopted as roadside pavement without taking up any additional space for $NO_x$ removal.

## 5.3.2 $NO_x$ Removal Performance of Photocatalytic Blocks

### 5.3.2.1 Relationship Between UV Irradiation and $NO_x$ Removal Performance

Figure 5.3 shows the relationship between ultraviolet rays (UV) irradiance and $NO_x$ removal rate of a photocatalytic interlocking block, measured by the flow-through method [34]. At a UV irradiance of 0.0 W/m², the photocatalytic block hardly removes the $NO_x$. It removes $NO_x$ under the UV irradiance of more than 1.0 W/m². The higher the UV irradiance is, the higher the $NO_x$ removal rate is. The $NO_x$ removal rate is almost fixed under the UV irradiance over 6.0 W/m². UV radiation found in sunlight is about 4% of all light energy. The UV irradiance by the sunlight is 20–30 W/m² during the fine weather of the Japanese summer. The UV irradiance in the outdoors is about 1 W/m² on days of cloudy winter weather. Therefore, the photocatalytic block can remove $NO_x$ on the roadside.

### 5.3.2.2 NO Supply Concentration and $NO_x$ Removal Performance

Figure 5.4 represents the relationship between NO supply concentration and $NO_x$ removal rate of a photocatalytic interlocking block, measured by the flow-through

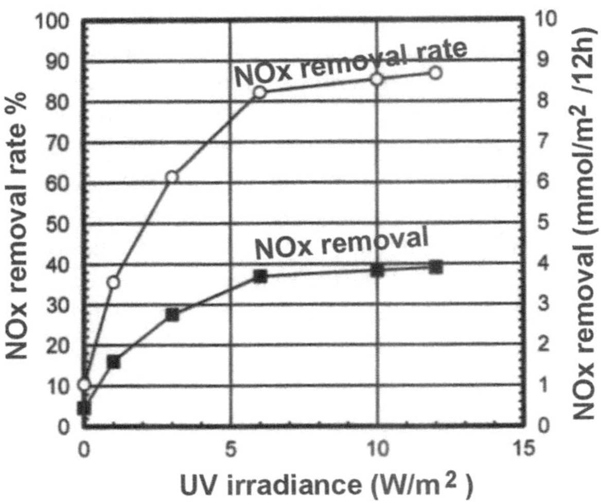


**Fig. 5.3** Relationship between UV irradiance and $NO_x$ removal of photocatalytic interlocking block at NO supply concentration of 1 ppm

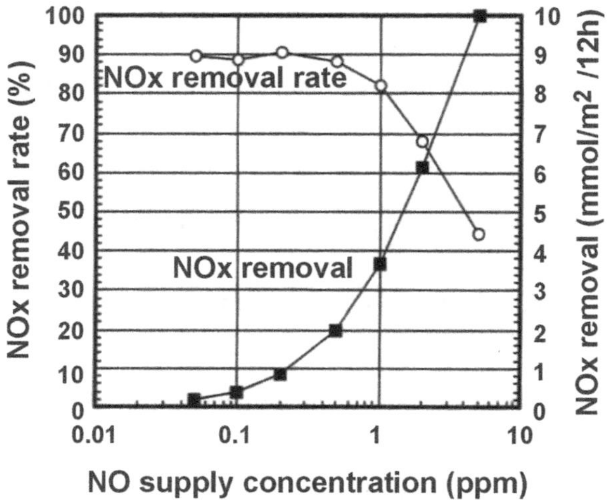


**Fig. 5.4** Relationship between NO supply concentration and $NO_x$ removal at UV irradiance of 6 W/m$^2$

method [34]. The $NO_x$ removal rate indicates almost 80% or more in the N0 concentration range from 0.05 to 1.0 ppm. Generally, the $NO_x$ concentration measured is 1.0 ppm or less on roadsides. Therefore, we need the technology that removes low-concentration $NO_x$ in the air. The photocatalyst effectively removes low-concentration $NO_x$ in the air. The $NO_x$ removal is proportional to the NO supply concentration under 1.0 ppm in the air.

#### 5.3.2.3 Long-Term $NO_x$ Removal Performance

The photocatalytic interlocking blocks were set up outdoors. Sets of two blocks were collected regularly. The long-term $NO_x$ removal performance was measured using a ventilation type-measuring apparatus. The $NO_x$ removal performance degradates in several months in the early ages as shown in Fig. 5.5 [34]. Afterwards, $NO_x$ removal performance is almost constant at 2 mmol/m$^2$/12 h. The main causes of the reduced $NO_x$ removal performance are attributed to the decreased effective $NO_x$ adsorption area and the shutout of light due to the adhesion of fine dust and organisms. However, if a periodical road cleaning is carried out, the activity is almost completely recovered. Sometimes, the cleaning action of rain is already sufficient.

#### 5.3.2.4 Examinations in an Urban Area Near Roads

There are two methods that examine the effect of $NO_x$ removal in an urban area near roads. One is a method of analyzing the nitrate and nitrite ions accumulated.

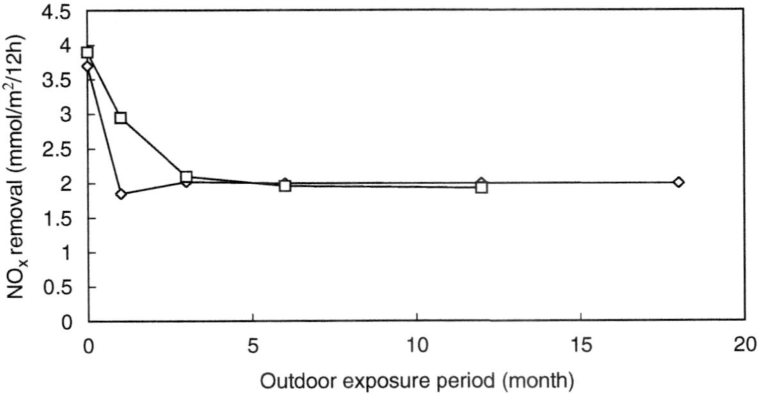


**Fig. 5.5** Long-term $NO_x$ removal performance of photocatalytic interlocking blocks

**Fig. 5.6** $NO_x$ concentration (average for each 1 week) at Points A and B vs. elapsed time

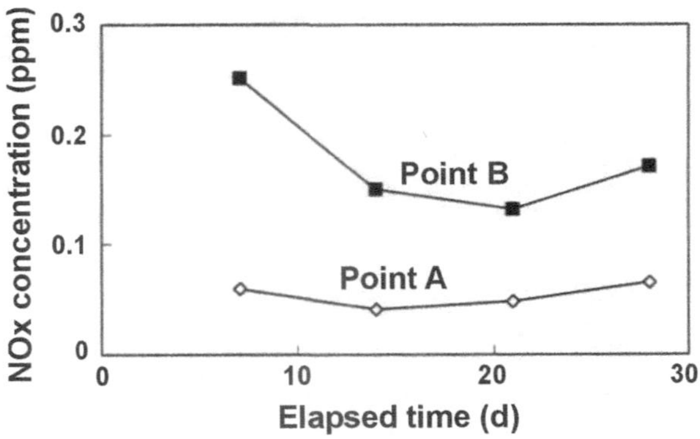


Another is a method of measuring a decrease in the $NO_x$ concentration in the air [21]. The method that examines the accumulated nitrate ions is an effective measure because it is easily enforceable. However, it has the fault that the nitrate ions cannot completely be eluted.

The photocatalytic blocks were set up in the various points, and exposed for a prescribed period (1 month).

Figures 5.6 and 5.7 show $NO_x$ concentration at Points A and B and the accumulated $NO_x$ removal of photocatalytic interlocking blocks for 1 month, respectively [34]. Point B is located near to the roadway compared to Point A. As a matter of course, the $NO_x$ concentration at Point B is much higher than that at Point A. The accumulated $NO_x$ removal at Point A is 5.7 mmol/m². At Point B near the roadway, the $NO_x$ removal is 12.3 mmol/m². The accumulated $NO_x$ removal is increasing after 3 weeks, and the $NO_x$ removal performance is hold.

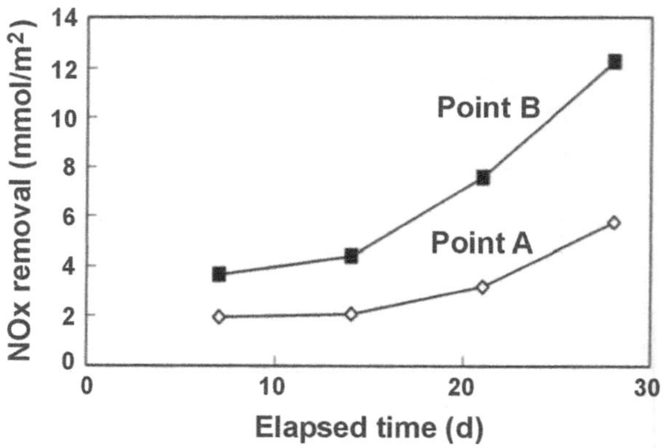


**Fig. 5.7** Relationship between accumulated $NO_x$ removal and elapsed time

**Fig. 5.8** Concrete blocks pavement in Kawasaki City, Japan in1999

## *5.3.3 Examples of Applications*

The photocatalytic blocks are mainly used as pavements (side-walks) and roadways. The pavements shown in Figs. 5.8, 5.9 and 5.10 are typical examples. Over one million m$^2$ of pavements using photocatalytic blocks were constructed worldwide up to 2007. Up to 25,000 m$^2$ of pavements using photocatalytic blocks were constructed in Japan up to 2006.

**Fig. 5.9** Concrete blocks pavement in Saitama City, Japan in 2000–2004

**Fig. 5.10** Concrete blocks pavements in Bergamo, Italy in 2006

**Fig. 5.11** Separate parking lanes at the Leien of Antwerp with photocatalytic pavement blocks

Over the years, different pilot projects were carried out in several countries. For example, photocatalytic materials were applied in the wearing course of pavement blocks in Antwerp [4]. 10,000 m² on the parking lanes next to the Leien of Antwerp, an inner city ring road, were placed with photocatalytic pavement blocks in 2005. The efficiency is determined over time in the laboratory by measuring the $NO_x$-reduction. A view on the Leien is given in Fig. 5.11.

In Northern Italy, several concrete pavements were built by using photocatalytic interlocking blocks near commercial zones and in residential areas [11]. In the centre of Bergamo (Italy), a heavy-traffic street has recently been renovated using 12,000 m² of photocatalytic paving blocks in September 2006, as shown in Fig. 5.10. Monitoring data have demonstrated the efficiency, quantified in almost 40% of $NO_x$ abatement even in winter conditions [21]. Other construction sites were completed in different European countries, e.g., Paris (France), 'Portes de Vanves', Fulda (Germany).

## 5.4 Photocatalytic Pavements by Coating of Photocatalytic Polymer-Modified Pastes

The above-mentioned photocatalytic blocks make the pavements which perform to purify the air. On the other hand, there is a process in which photocatalytic materials are directly coated on the surfaces of drainage pavements and develop a photocatalytic function to purify the air. The process was developed by Taiheiyo Cement Corporation in Japan in 1999 [26]. Polymer-modified pastes or slurries using a polyacrylic ester emulsion with $TiO_2$ photocatalyst are used as photocatalytic materials [40]. The drainage pavements are executed by using gap-graded (or open-graded) asphalt concretes. In general, coating process is applied by a spraying or pouring method. The photocatalytic pavements constructed by such execution processes can effectively decompose car exhaust gases with $NO_x$ on road surfaces. Figure 5.12

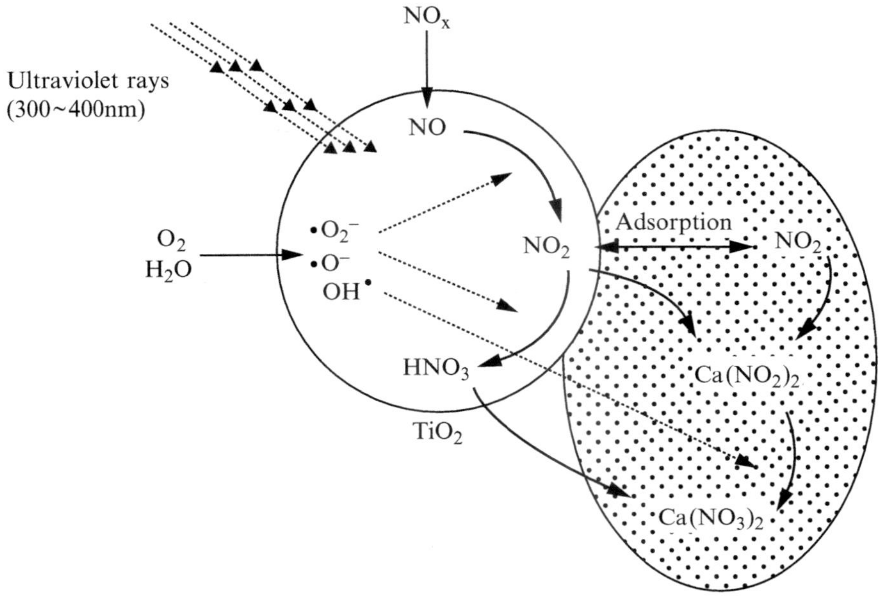


**Fig. 5.12** $NO_x$ purification mechanism of photocatalytic pavements, executed by coating of photocatalytic polymer-modified pastes or slurries

illustrates the $NO_x$ purification mechanism of the photocatalytic pavements, executed by the coating of photocatalytic polymer-modified pastes or slurries [40]. $NO_x$ is oxidized by the active oxygens generated by light irradiation, and finally fixed as nitrate ions ($NO_3^-$) or nitric acid ($HNO_3$), which react with cement hydrates to form calcium nitrate [$Ca(NO_3)$]. The calcium nitrate accumulated on the coating surfaces can easily be washed out by rainfall. Therefore, the $NO_x$ purification performance of the photocatalytic pavements can be maintained for a long time. Figure 5.13 represents the effects of UV irradiation time and $NO_x$ type on the $NO_x$ purification performance of a photocatalytic polymer-modified paste or slurry coating in a laboratory test [40]. Regardless of the $NO_x$ type, the $NO_x$ concentration is decreased with increasing UV lamp irradiation time, and becomes nearly 0 ppm within a UV lamp irradiation time of 20 min. It is found from this figure that the coating develops its high $NO_x$ purification performance within a short time. The photocatalytic pavements can prevent air pollution by semipermanently holding their air purification function, have high abrasion resistance to car traffic, and have water permeability and sound-absorbing property. Since 1999, the photocatalytic polymer-modified paste- or slurry-coated pavements have been put to practical use in Japan, and 27,400 $m^2$ of the pavements have been executed up to 2007.

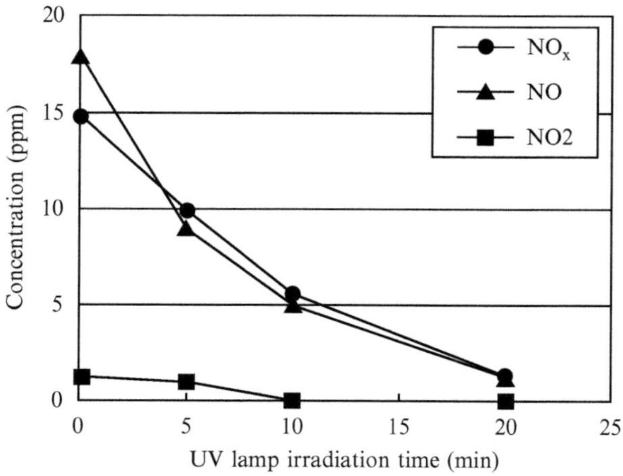


**Fig. 5.13** Effects of UV lamp irradiation time and $NO_x$ type on $NO_x$ purification performance of photocatalytic polymer-modified paste or slurry coating

**Fig. 5.14** Coating equipment with a mixer and spray applicator for polymer-modified pastes or slurries on a truck (Courtesy of Taiheiyo Cement Corporation)

**Fig. 5.15** Drainage pavement and railing wall coated with photocatalytic polyacrylic ester-modified pastes in an expressway (Courtesy of Taiheiyo Cement Corporation)

Figure 5.14 shows a coating equipment with a mixer and spray applicator for the photocatalytic polymer-modified pastes or slurries for spraying work. Figure 5.15 represents the drainage pavement and railing wall coated with the photocatalytic polyacrylic ester-modified pastes in an expressway.

# Chapter 6
# Standardization of Testing Methods for Construction Materials with $TiO_2$ Photocatalyst

**Kenji Motohashi, Frank Dehn, and Yoshihiko Ohama**

## 6.1 Introduction

In order to get uniformity for the determination and verification of test results for photocatalytic construction materials, it is essential to define performance-based attributes and test methods. This chapter summarizes the current state-of-the-art on the existing standards, and shows an example for a performance-based procedure which covers the removal of nitric oxides ($NO_x$).

Such an approach will help to compare the performance of different photocatalytic construction materials, and provide not only information on a scientific level but also for potential consumers of such photocatalytic products.

$TiO_2$ photocatalyst has practically been applied to the field of construction materials for the following purposes:

1. Self-cleaning effects
   $TiO_2$ photocatalyst has been applied on building finishing materials to impart self-cleaning capability to them. Not only for inorganic substrates such as ceramic tiles, natural stone, glass, inorganic material boards, etc. but also for organic finishing materials such as paints, plastics sheets, membranes, etc., the photocatalyst has been coated after the application of a barrier layer which prevents organic substrates from photochemical degradation due to photocatalysis.
2. Removal of nitric oxide
   $TiO_2$ photocatalyst has been applied to roads and related facilities for trials. The photocatalyst can be expected to remove nitric oxide which is one of harmful contaminants in the traffic exhaust gases.

K. Motohashi
Shibaura Institute of Technology, College of Engineering,
Department of Architecture and Building Engineering, Japan

F. Dehn (✉)
University of Leipzig, Germany
e-mail: dehn@mfpa-leipzig.de

Y. Ohama
Nihon University, Tokyo, Japan

Y. Ohama and D. Van Gemert (eds.), *Application of Titanium Dioxide Photocatalysis to Construction Materials*, RILEM State-of-the-Art Reports 5,
DOI 10.1007/978-94-007-1297-3_6, © RILEM 2011

3. Antibacterial effects and deodorization effects
   Interior tiles, interior wallpapers, interior finishing paints and floorings have sometimes been treated by $TiO_2$ photocatalyst for the purpose of expecting antibacterial effects and deodorization effects to volatile organic compounds. These interior finishing materials are mostly used in hospitals, nursing homes, schools, etc. Of course, there are arguments on the effectiveness of such interior finishing materials under normal indoor illumination.
4. Cooling effects
   Application of $TiO_2$ photocatalyst to the envelopes of buildings and road facilities has been studied in order to control heat island phenomena and tropical night phenomena by the cooling effects due to superhydrophilicity.
5. Water treatment and soil treatment
   To decompose very dilute contaminants such as environmental endocrine disrupter, dioxin, PCB, etc. in water and soil, new treatment methods with the application of $TiO_2$ photocatalyst have been investigated in Japan.

Among various application fields of $TiO_2$ photocatalysis in construction, building finishing materials with self-cleaning ability are most popular. Some prefabricated house manufacturers supplied new type of houses with photocatalyst-coated tile finishing or external finishing boards on the market a few years ago.

## 6.2 The Existing Standards

Table 6.1 shows an overview on the present status of different international and national standards which are published or under preparation. The standards cover the performance attributes mentioned in Sect. 6.1.

**Table 6.1** Overview of international and national standards related to $TiO_2$ photocatalysis as of May 30, 2008

| Performance attributes | Principle of test method | National standard (e.g., Japan and Italy) | International standard |
|---|---|---|---|
| Air purification effect | Nitric oxide removal | JIS R 1701–1:2004<br>UNI 1247:2007 | ISO 22197–1:2007 |
| | Removal Volatile Organic Compound | UNI 11238–1:2007<br>11238–2:2007 | |
| | Acetaldehyde removal | JIS R 1701–2: 2008 | *ISO/CD 22197-2* |
| | Toluene removal | JIS R 1701–3: 2008 | *ISO/CD 22197-3* |
| Water purifycation effect | Active oxygen-forming | JIS R 1704:2007 | – |
| Self-cleaning effect | Water contact angle change | JIS R 1703–1:2007 | *ISO/CD 27448-1* |
| | Methylene blue decomposition | JIS R 1703–2:2007 | *ISO/CD 10678* |
| | Rhodamine | UNI 11259:2008 | |
| Biocidal effect | Antibacterial activity | JIS R 1702:2006 | *ISO/DIS 27447* |
| | Antifungal activity | JIS R 1705:2008 | – |
| – | Light source for test under UV irradiation | JIS R 1709:2007 | – |

Note: Standards *in italics* are under preparation

**Table 6.2** International, Japanese and Italian standards related to $TiO_2$ photocatalysis

| Standard number and year | Title of standard |
|---|---|
| ISO 22197–1:2007 | Fine ceramics (advanced ceramics, advanced technical ceramics) – Test method for air-purification performance of semiconducting photocatalytic materials – Part 1: Removal of nitric oxide |
| JIS R 1701–1:2004 | Fine ceramics (advanced ceramics, advanced technical ceramics) – Test method for air purification performance of photocatalytic materials – Part 1: Removal of nitric oxide |
| JIS R 1701–2:2008 | Fine ceramics (advanced ceramics, advanced technical ceramics) – Test method for air purification performance of photocatalytic materials – Part 2: Removal of acetaldehyde |
| JIS R 1701–3:2008 | Fine ceramics (advanced ceramics, advanced technical ceramics) – Test method for air purification performance of photocatalytic materials – Part 3: Removal of toluene |
| JIS R 1702:2006 | Fine ceramics (advanced ceramics, advanced technical ceramics) – Test method for antibacterial activity of photocatalytic products under photoirradiation and efficacy |
| JIS R 1703–1:2007 | Fine ceramics (advanced ceramics, advanced technical ceramics) – Test method for self-cleaning performance of photocatalytic materials – Part 1: Measurement of water contact angle |
| JIS R 1703–2:2007 | Fine ceramics (advanced ceramics, advanced technical ceramics) – Test method for self-cleaning performance of photocatalytic materials – Part 1: Decomposition of wet methylene blue |
| JIS R 1704:2007 | Fine ceramics (advanced ceramics, advanced technical ceramics) – Test method for water-purification performance of photocatalytic materials by measurement of forming ability of active oxygen |
| JIS R 1705:2008 | Fine ceramics (advanced ceramics, advanced technical ceramics) – Test method for antifungal activity of photocatalytic products under photoirradiation |
| JIS R 1709:2007 | Light source for test of photocatalytic materials used under ultraviolet |
| UNI 11247: 2007 | Determination of the catalytic degradation of nitrogen oxides in air by photocatalytic inorganic materials |

One ISO and nine Japanese Industrial Standards (JISs) and five Italian Standards (UNIs) related to $TiO_2$ photocatalysis have been published recently as seen in Tables 6.2 and 6.3. Standardization of evaluation test methods for various performance of photocatalytic materials has been intensively pushed forward.

Figure 6.1 shows a scheme of the test equipment in JIS R 1701–1:2004 for evaluation of air-purification performance of photocatalytic materials. This test equipment is similar to that in ISO 22197–1:2007, which was standardized on the basis of JIS R 1701–1:2004.

Figure 6.2 represents an example of measuring water contact angle according to JIS R 1703–1:2007. The surface of a test piece is firstly coated with oleic acid, and is irradiated by ultraviolet light. The water contact angle of the surface is measured

**Table 6.3** Italian standards related to $TiO_2$ photocatalysis

| Standard number and year | Title of standard |
|---|---|
| UNI 11247:2007 | Determination of the catalytic degradation of nitrogen oxides in air by photocatalytic inorganic materials |
| UNI 11238–1:2007 | Determination of the catalytic degradation of organic micropollutants in air – Part 1: Photocatalytic cementitious materials |
| UNI 11238–2:2007 | Determination of the catalytic degradation of organic micropollutants in air – Part 2: Photocatalytic ceramic materials for use in construction |
| UNI 11259:2008 | Determination of the photocatalytic activity of hydraulic binders – Rhodamine test method |

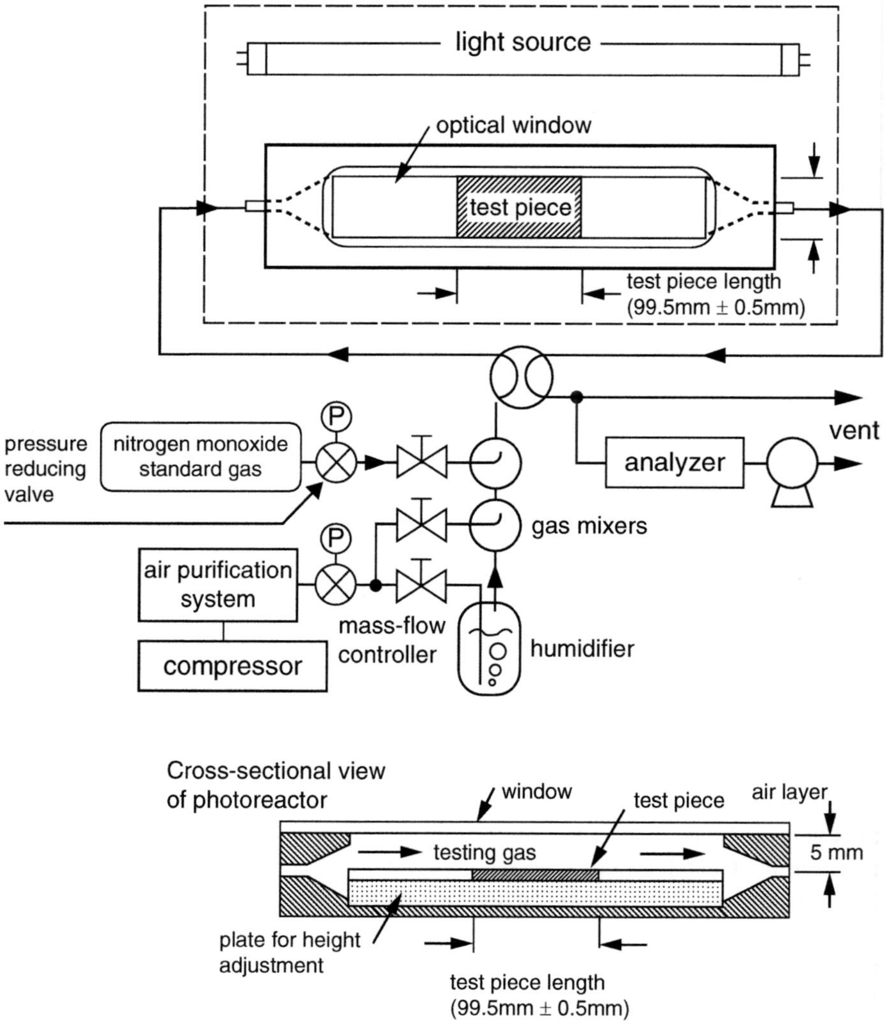


**Fig. 6.1** Scheme of test equipment for evaluating air purification performance (JIS R 1701–1:2004)

periodically, and the "critical contact angle" of the surface will be finally determined in accordance with JIS R 1703–1:2007.

Figure 6.3 illustrates a schematic of an experimental method for the evaluation of wet methylene blue decomposition.

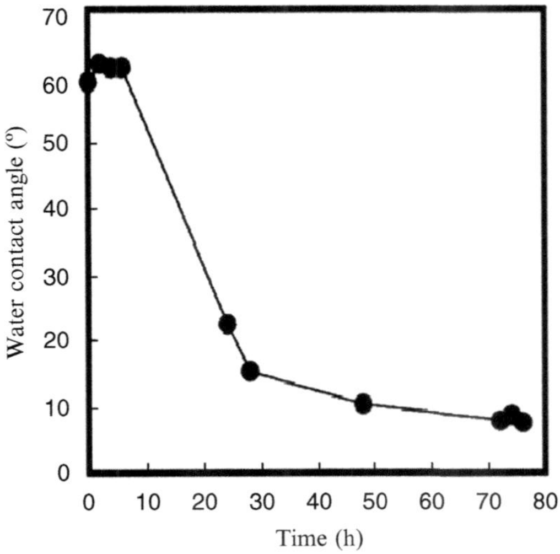


Example in which pretreatment (oleic acid application method) of test piece is performed by "dip".

**Fig. 6.2** Example of measurement of water contact angles (JIS R 1703–1:2007)

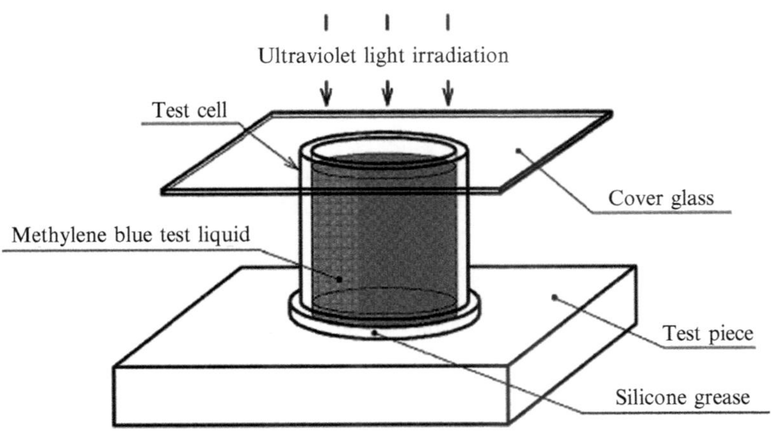


**Fig. 6.3** Schematic of experimental method for evaluation of wet methylene blue decomposition (JIS R 1703–2:2007)

# Chapter 7
# Conclusions

**Yoshihiko Ohama and Dionys Van Gemert**

Since the early applications in 1972 of photocatalysis by titanium dioxide by Fujishima and Honda [15], a rapid progress has taken place. The technology of $TiO_2$ photocatalyst or photocatalysis has become very attractive in the development of construction materials and construction techniques for self-cleaning, air or water purification, and antibacterial functions. No further chemicals have to be used, only sunlight and rainwater. Photocatalyst technology is being applied on ceramic, glass and cementitious surfaces, as well as in paints. Exterior as well as interior applications exploit the benefits of $TiO_2$ photocatalysis. Application of photocatalysis to the envelopes of buildings, road facilities and tunnels reduces air pollution, and thus contributes to an enhanced control of heat island phenomena in densely populated cities or heavy traffic areas. Whereas the effects of photocatalysis increase with the surface area exposed to sunlight, roofs and roofing felts are a rapidly increasing field. Novel functions of photocatalysis are expected through the development of visible light-sensitive $TiO_2$ photocatalyst, hydrophobic $TiO_2$ photocatalyst turning into highly hydrophilic under irradiation, … . Life cycle analysis shows the positive balance of $TiO_2$ photocatalyst production and application.

Now that $TiO_2$ photocatalysis application is rapidly growing, the need for standardization becomes urgent. As Japan is the cradle of $TiO_2$ photocatalysis, standards became available in Japan first. Taking benefit of the Japanese experiences, international standards should be developed and published without delay.

A number of aspects of $TiO_2$ photocatalysis still need further concern and research. A nonexhaustive list of elements for further research comprises the study of the influence of relative humidity on photocatalytic performance and efficiency, the influence on the health of $TiO_2$-nanoparticles eventually released in the air, and the stability of $TiO_2$ with time.

Y. Ohama (✉)
Nihon University, Tokyo, Japan
e-mail: ohama@bh.wakwak.com

D. Van Gemert
Katholieke Universiteit Leuven, Heverlee, Belgium

Y. Ohama and D. Van Gemert (eds.), *Application of Titanium Dioxide Photocatalysis to Construction Materials*, RILEM State-of-the-Art Reports 5,
DOI 10.1007/978-94-007-1297-3_7, © RILEM 2011

# References

1. Adamson AW (1982) Physical chemistry of surfaces, 4th edn. Wiley, New York
2. Asahi R, Morikawa T, Ohwaki T, Aoki K, Taga Y (2001) Visible-light photocatalysis in nitrogen-doped titanium oxides. Science 293(5528):269–271
3. Baglioni P, Cassar L (2007) Proceedings of the international RILEM symposium on photocatalysis, environment and construction materials – TDP 2007. RILEM Proceedings PRO 55. RILEM Publications, Bagneux
4. Beeldens A, Redant K (2006) Air purification and self-cleaning materials: photocatalytic pavement blocks on the "Leien" of Antwerp. In: Proceedings of the 10th international symposium on concrete roads, Febelcem-Brussels, Brussels, Sept 2006
5. Benedix R, Dehn F, Quass J, Orgass M (2000) Application of titanium dioxide photocatalysis to create self-cleaning building materials. Leipzig Annual Civil Engineering Report LACER No.5. Universität Leipzig, Leipzig, Germany, pp 157–167
6. Blöß SP, Elfenthal L (2007) Doped titanium dioxide as photocatalyst for UV and visible light. In: Baglioni P, Cassar L (eds.) Proceedings of the international RILEM symposium on photocatalysis, environment and construction materials – TDP 2007, Florence. RILEM Publications, Bagneux, Oct 2007, pp 31–38
7. Bygott C, Maltby J, Stratton J, McIntyre R (2007) Photocatalytic coatings for the construction industry. In: Baglioni P, Cassar L (eds.) Proceedings of the international RILEM symposium on photocatalysis, environment and construction materials – TDP200, Florence. RILEM Publications, Bagneux, Oct 2007, pp 251–258
8. Calgon (1996) The AOT handbook. Calgon Carbon, oxidation technologies. Calgon, Ontario
9. Cassar L (2004) Photocatalysis of cementitious materials: clean buildings and clean air. MRS Bull 29(5):328–331
10. Cassar L, Pepe C, Tognon G, Guerrini GL, Amadelli R (2003) White cement for architectural concrete, possessing photocatalytic properties. In: Proceedings of the 11th international congress on the chemistry of cement, vol 4. The Cement and Concrete Institute of South Africa, Durban, pp 2012–2021
11. Cassar L, Beeldens A, Pimpinelli N, Guerrini G (2007) Photocatalysis of cementitious materials. In: Baglioni P, Cassar L (eds.) Proceedings of the international RILEM symposium on photocatalysis, environment and construction materials – TDP 2007, Florence. RILEM Publications, Bagneux, pp 131–143
12. De Muynck W, Maury A, De Belie N, De Bock J, Verstraete W (2008) Evaluation of antifouling strategies on aerated concrete by means of an accelerated algal growth test. In: Alexander MG, Beushausen HD, Dehn F, Moyo P (eds.) Proceedings of the 2nd international conference on concrete repair, rehabilitation and retrofitting (ICCRRR), Cape Town. CRC Press, Boca Raton, Nov 2008, pp 131–132, in abstract book; 253–259 on CD-ROM. ISBN: 978-0-415-46850-3

Y. Ohama and D. Van Gemert (eds.), *Application of Titanium Dioxide Photocatalysis to Construction Materials*, RILEM State-of-the-Art Reports 5,
DOI 10.1007/978-94-007-1297-3, © RILEM 2011

13. EPA (1985) Glossary of terms used in photochemistry. European Photochemistry Association (EPA) Newsletter 25. Loughborough University, Leicestershire, p 13
14. Fox MA (1988) Photocatalytic oxidation of organic substrates. In: Schiavello M (ed.) Proceedings of the NATO ASI international conference on photocatalysis and environment: trends and applications, Palermo. Kluwer, Dordrecht, 1987, pp 445–467
15. Fujishima A, Honda H (1972) Electrochemical photolysis of water at a semiconductor electrode. Nature 238(5358):37–38
16. Fujishima A, Hashimoto K, Watanabe T (1999) $TiO_2$ photocatalysis – fundamentals and applications. BKC, Tokyo
17. Fujishima A, Rao TN, Tryck DA (2000) Titanium dioxide photocatalysis. J Photochem Photobiol C 1:1–21, Photochemistry Review
18. Gaylarde Ch, Gaylarde P (2005) A comparative study of the major microbial biomass of biofilms on exteriors of buildings in Europe and Latin America. Int Biodeter Biodegr 55:131–139
19. Gelover S, Gómez LA, Reyes K, Leal MT (2006) A practical demonstration of water disinfection using $TiO_2$ films and sunlight. Water Res 40:3274–3280
20. Glaze WH (1987) Drinking–water treatment with ozone. Environ Sci Technol 21:224–230
21. Guerrini G, Peccati L (2007) Photocatalytic cementitious roads for depollution. In: Baglioni P, Cassar L (eds.) Proceedings of the international RILEM symposium on photocatalysis, environment and construction materials – TDP 2007, Florence. RILEM Publications, Bagneux, Oct 2007, pp 179–186
22. Harrison WA (1980) Electronic structure and the properties of solids. W. H. Freeman & Co., New York
23. Hashimoto K (2007) TiO2 photocatalysts towards novel building materials. In: Baglioni P, Cassar L (eds.) Proceedings of the international RILEM symposium on photocatalysis, environment and construction materials – TDP 2007, Florence. RILEM Publications, Bagneux, Oct 2007, pp 3–8
24. Hashimoto K, Fujishima A (2003) Zukai hikari-shokubai-no-subete. Kogyochosakai, Tokyo, in Japanese
25. Herrmann JM, Duchamp C, Karkmaz M, Hoai BT, Lachheb H, Puzenat E, Guillard C (2007) Environmental green chemistry as defined by photocatalysis. J Hazard Mater 146:624–629
26. Ishimori M (2000), Photo-road with function to remove car exhaust gas by optical catalyst, Japan Cement Association of Cement & Concrete 639:18–23 (in Japanese)
27. Italcementi (2005) TX millennium photocatalytic binders (Technical report). Italcementi, Bergamo
28. King DA, Woodruff WP (eds.) (1982) The chemical physics of solid surfaces and heterogeneous catalysis, vol 1–4. Elsevier, Amsterdam
29. Kisch H (1989) What is photocatalysis? In: Serpone N, Pelizzetti E (eds.) Photocatalysis: fundamentals and applications. Wiley Interscience, New York
30. Kurth JC, Giannantonio DJ, Allain F, Sobecky P, Kurtis KE (2007) Mitigating biofilm growth through the modification on concrete design and practice. In: Baglioni P, Cassar L (eds.) Proceedings of the international RILEM symposium on photocatalysis, environment and construction materials – TDP 2007, Florence. RILEM Publications, Bagneux, Oct 2007, pp 195–202
31. Maggos T, Plassais A, Bartzis J, Vasilakos C, Moussiopoulos N, Bonafous L (2005) Photocatalytic degradation of $NO_x$ in a pilot street canyon configuration using $TiO_2$-mortar panels. In: Sokhi RS, Neophytou M (eds.) Proceedings of the 5th international conference on urban air quality, PICADA section, Valencia. Springer, Amsterdam, Mar 2005
32. Marinoni N, Pellizon Birelli M, Rostagno Ch, Pavese A (2003) The effects of atmospheric multipollutants on modern concrete. Atmos Environ 37:4701–4712
33. Murata Y (2000) A new photocatalytic technology for the atmospheric $NO_x$ removal. J Soc Inorg Mater 7:323–330 (in Japanese)

34. Murata Y, Tobinai K (2002) Influence of various factors on $NO_x$ removal performance of permeable interlocking block based on photocatalysis. J Struct Construct Eng 555:9–15, Transaction of Architectural Institute of Japan (in Japanese)
35. O´Keeffe M, Navrotsky A (1981) Structure and bonding in crystals. Academic, New York
36. Pappas SP, Fischer RM (1974) Photo-chemistry of pigments—studies on the mechanism of chalking. J Paint Technol 46(599):65–72
37. Peller JR, Whitman RL, Griffith S, Harris P, Peller C, Scalzitti J (2007) $TiO_2$ as a photocatalyst for control of the aquatic invasive alga, Cladophora, under natural and artificial light. J Photochem Photobiol A 186:212–217
38. Plotnikow J (1936) Allgemeine Photochemie, 2nd edn. Walter de Gruyter, West Berlin
39. Sakai N, Fujishima A, Watanabe T, Hashimoto K (2003) Quantitative evaluation of the photoinduced hydrophilic conversion properties of $TiO_2$ thin film surfaces by the reciprocal of contact angle. J Phys Chem B 107(4):1028–1035
40. Sano K, Kaneda Y, Ishimori M (2002) The $NO_x$ purification mechanism of photocatalyst cement (ST-coat). J Res Taiheiyo Cement Corp 142:63–70, Taiheiyo Cement Kenkyu Hokoku (in Japanese)
41. Serpone N, Pelizzetti E (1989) Photocatalysis: fundamentals and applications. Wiley-Interscience, Amsterdam
42. Shimohigoshi M, Saeki Y (2004) Development of photocatalyst tile and commercial production. In: Kashino N, Ohama Y (eds.) Proceedings of the RILEM international symposium on environment-conscious materials and systems for sustainable development, Koriyama, Sept. 2004. RILEM proceedings PRO 41. RILEM Publications, Bagneux, pp 35–40
43. Sunada K, Watanabe T, Hashimoto K (2003) Studies on photokilling bacteria on $TiO_2$ thin film. J Photochem Photobiol A 156:227–233
44. Taoda H (2004) The newest trends in development of photocatalytic materials. Eng Mater 52(11):32–37, Kogyo-Zairyo (in Japanese)
45. Tilley JD (1987) Defect crystal chemistry. Blackie, London
46. Vanselow R, Howe R (eds.) (1979) Chemistry and physics of solid surfaces, vol 1–6. Springer, Berlin
47. Wang R, Hashimoto K, Fujishima A, Chikuni M, Kojima E, Kitamura A, Shimohigashi M, Watanabe T (1997) Light-induced amphiphilic surfaces. Nature 388(6641):431–432
48. Watanabe T, Nakajima A, Wang R, Minabe M, Koizumi S, Fujishima A, Hashimoto K (1999) Photocatalytic activity and photoinduced hydrophilicity of titanium dioxide coated glass. J Thin Solid Films 351:260–263
49. Wubbels G (1983) Catalysis of photochemical reactions. Acc Chem Res 16:285–292

GPSR Compliance

*The European Union's (EU) General Product Safety Regulation (GPSR) is a set of rules that requires consumer products to be safe and our obligations to ensure this.*

*If you have any concerns about our products, you can contact us on ProductSafety@springernature.com*

In case Publisher is established outside the EU, the EU authorized representative is:

Springer Nature Customer Service Center GmbH
Europaplatz 3
69115 Heidelberg, Germany

**Batch number: 10607224**

Printed by Printforce, the Netherlands